Children, Youth and the City

More than half of the global and around 80 per cent of the Western population grow up in cities. This text provides a vivid picture of children and youth in the city, how they make sense of it and how they appropriate it through their social actions.

Considering the causes and forms of social inequalities in relation to class, gender, ethnicity, sexuality, ability and geographical location, this book discusses specific issues such as poverty, homelessness and work. Each chapter draws on examples from both the Global North and South, and throughout the chapters, the book:

- contrasts experiences of growing up in the city
- discusses how social inequalities, together with societal perceptions of childhood and youth, shape experiences of growing up in cities for different young people
- examines how young people appropriate the city through social and cultural practices
- considers contemporary movements towards the role of children and youth in planning processes.

Children, Youth and the City argues that young people must be recognised as urban social agents in their own right. This informative book deals with complex theoretical arguments and relates key ideas to this topical subject in a clear and coherent manner. The text is enlivened throughout with global case studies, photographs, discussion questions, suggested reading and websites. It is an excellent resource for students of Human Geography, Urban Studies and Childhood Studies.

Kathrin Hörschelmann is Senior Lecturer in Human Geography at the University of Durham, UK.

Lorraine van Blerk is Senior Lecturer in Human Geography at the University of Dundee, UK.

Routledge critical introductions to urbanism and the city

The series is designed to allow undergraduate readers to make sense of, and find a critical way into, urbanism. It will:

- cover a broad range of themes
- introduce key ideas and sources
- allow the author to articulate her/his own position.
- introduce complex arguments clearly and accessibly
- bridge disciplines, and theory and practice
- be affordable and well designed

The series covers social, political, economic, cultural and spatial concerns. It will appeal to students in architecture, cultural studies, geography, popular culture, sociology, urban studies, urban planning. It will be transdisciplinary. Firmly situated in the present, it also introduces material from the cities of modernity and postmodernity.

Published:

Cities and Consumption – Mark Jayne
Cities and Cultures – Malcolm Miles
Cities and Nature – Lisa Benton-Short and John Rennie Short
Cities and Economies – Yeong-Hyun Kim and John Rennie Short
Cities and Cinema – Barbara Mennel
Cities and Gender – Helen Jarvis with Paula Kantor & Jonathan Cloke
Cities and Design – Paul L. Knox
Cities, Politics and Power – Simon Parker
Cities and Sexualities – Phil Hubbard
Children, Youth and the City – Kathrin Hörshelmann and Lorraine van Blerk

Forthcoming:

Cities and Climate Change – Harriet A. Bulkeley
Cities and Photography – Jane Tormey
Cities, Risk and Disaster – Christine Wamsler

Children, Youth and the City

Kathrin Hörschelmann and Lorraine van Blerk

LONDON AND NEW YORK

First published 2012
by Routledge
2 Park Square, Milton Park, Abingdon, Oxon OX14 4RN

Simultaneously published in the USA and Canada
by Routledge
711 Third Avenue, New York, NY 10017

Routledge is an imprint of the Taylor & Francis Group, an informa business

British Library Cataloguing in Publication Data
A catalogue record for this book is available from the British Library

Library of Congress Cataloging-in-Publication Data
Hörschelmann, Kathrin, 1971-
Children, youth and the city / Kathrin Hörschelmann and Lorraine van Blerk.
p. cm.
Includes bibliographical references and index.
1. Children–Social conditions. 2. Youth–Social conditions. 3. City and town life. 4. Sociology, Urban. I. Van Blerk, Lorraine II. Title.
HQ767.9.H67 2011
305.2309173'2–dc23

ISBN: 978-0-415-37693-8 (hbk)
ISBN: 978-0-415-37692-1 (pbk)
ISBN: 978-0-203-96756-0 (ebk)

Typeset in Times New Roman by
GreenGate Publishing Services, Tonbridge, Kent

Printed and bound in Great Britain by
CPI Antony Rowe, Chippenham, Wiltshire

To our children

Contents

Figures

Boxes

Acknowledgements

This book would never have come about without the great dedication and engagement of the children and youth who have participated in a number of research projects that we have conducted in Germany, the UK, Pakistan and East and Southern Africa. Much of the inspiration for the book has come from our research with children and youth in these places and it has been their reflections on their urban lives that have opened our eyes to many of the issues currently faced by young people in different urban settings around the globe. We would like to thank all of them and the many dedicated adult carers who have worked with us on one of these projects over the last ten years for sharing their knowledge and reflections with us.

We owe a great debt of gratitude to our editors at Routledge, especially Andrew Mould, Faye Leerink, Michael Jones and Zoe Kruze, who have given us immeasurable support and encouragement in seeing this project through to completion. We would like to thank them for their editorial guidance as well as their patience and understanding. Special thanks go to Malcolm Miles and John Rennie Short for inviting us to write this book and for their feedback on our initial ideas.

We have benefitted greatly from the generous and insightful comments of three anonymous reviewers and would like to express our gratitude for the great care that they took in giving feedback on the draft manuscript.

We thank our partners for their never-ending patience and support, and for sharing our belief in the need to create a better world for children and with children, in all parts of the world. Four of our children were born in the period between proposing and completing this book. We cannot begin to express how grateful we are for the joy and fulfilment that they have brought to our lives.

1 Introduction

In this chapter we will:

- **outline the key themes and rationales of the book**
- **sketch the main conceptual ideas that inform our approach to children, youth and the city**
- **summarise the main chapters of the book.**

More than half of the world's children live in cities today. It is here that they learn, play, work and contribute in a multitude of ways to the making of places. Far from being just a 'context' for these young people's lives, cities are significantly shaped by their activities and by the need to reflect their presence in the social organisation of life. Cities are *of* children and *of* youth in many ways. At the same time, what it means to be young is significantly shaped by the diverse ways in which cities and urban spaces are constructed and lived in the contemporary world. In order to understand how children, youth and the city relate, therefore, we need to consider both the manifold ways in which children and youth partake in the making of urban spaces, and the extent to which their experiences and practices correspond to wider social relations and the particular social and spatial construction of cities. In this book, we explore both sides of this relationship in order to show how and why young people's urban lives play out differently in diverse cities around the world, while being connected through global power relations and age-related hierarchies. We aim to show both how children and youth contribute to the making of urban spaces and how the particular structures, textures and materialities of cities inform how they live their diverse lives.

Children and youth play a key role in the making of urban spaces, both through their own activities and as a result of the adult-led management and disciplining of their needs and practices. In this book, we show how, even as children and youth are 'designed out' of many urban spaces, located in age-specific institutions and excluded from adult-dominated spaces, cities are intricately shaped by the presence and practices of young people. At the same time as the socio-spatial design and material texturing of cities limit the scope and shape of young people's actions, they provide resources that young people appropriate and (re) use in everyday life. Depending on their particular geographical location, they walk, cycle, travel by public transport or are driven to places of learning, work and play. They meet with friends on street corners, in urban squares or parks, go shopping, explore urban night-lives, volunteer, or simply 'hang out'. In many cities around the world, they run errands and work in the city to contribute to the household income. Some sell their own bodies on the street, beg, busk, deal or seek other legal and illegal ways of earning an income for their own survival and that of friends and family members. Others participate in urban life through subcultural activities that pose direct or indirect challenges to dominant and exclusionary uses of urban space.

Key approaches to children, youth and the city

There is now a rich body of work in geography and across the social sciences that explores and seeks to better understand these manifold contributions of young people to urban life (cf. McKendrick *et al.* 2000; Horton and Kraftl 2006). Scholars such as Colin Ward, David Lynch and Roger Hart produced nuanced and insightful accounts of children's experiences of urban space as early as the 1970s, critiquing the assumption that the city exits only 'for one particular kind of citizen: the adult, male, white collar, out of town car-user' (Ward 1978/1990: 25). Just over a decade later, critical commentaries by James (1990), Winchester (1991) and Sibley (1991) highlighted the continued absence of children from geographical research and paved the way for more sustained and critical engagements with young people's geographies that drew much of their inspiration from the so-called 'new social studies of childhood'. This new approach to the study of childhood questioned the universality of contemporary, Western definitions of age by demonstrating the extent to which generational categories such as childhood and youth are socially constructed and historically and culturally variable (James *et al.* 2001; Jenks 2005). It has drawn attention to the power relations and social hierarchies entailed in age categorisations (Qvortrup 2005; James and James 2004; Aitken 2001; Holloway and Valentine 2000), as well as examining how age intersects with other forms of social difference in a variety of global contexts. Moving away from essentialist understandings of age as defined by chronological and developmental stages, the new social studies of childhood and the emerging sub-discipline of children's geographies have adopted more *relational* approaches to age that recognise cultural and historical variability and the need to examine carefully the contexts in which age is defined and lived (Hopkins and Pain 2007). More recently, relational approaches to the geographies of children and youth have also begun to address questions of intergenerational exchange, recognising that definitions of age are related to how we understand generational positioning and that young people's lives are tightly linked to those of parents, grandparents, siblings and others (Evans and Becker 2009).

A further shift in geographical thinking that affects how we understand the urban lives of young people has been the call for greater attentiveness to the wide range of seemingly mundane, everyday practices and the materiality of urban lives (Horton and Kraftl 2006; Amin and Thrift 2002). This shift in thinking not only allows us to re-evaluate the frequently marginalised practices of children and youth, but also to broaden our understanding of the rhythms and textures of their lives, paying greater attention to the intersections between materiality, the body, memory, feelings, affects and practices, as well as those many minutiae that are an essential part of daily life but are difficult to capture in words. Such work challenges us to consider more carefully how different young people, through their bodies, encounter, engage with and produce urban space. In particular, it raises attention to some of the ruptures between urban spaces designed for particular uses and the bodies that (re)make them through diverse spatial practices.

In this book, we draw on these recent approaches to produce a relational account of the diverse ways in which childhood and youth are lived, experienced and imagined in different cities around the globe. We show the connections between social constructions of childhood, youth and the city, and the material performances, structures and textures of urban space. We consider the social, cultural and economic processes through which young people become marginalised residents in cities, how these age-related processes of marginalisation intersect with other forms of social and cultural difference, and how children and youth nonetheless use, produce and appropriate urban space. To understand young people's urban lives we need to consider all of these aspects as interrelated, and as culturally and historically contingent, as we seek to demonstrate throughout this book.

Outline of chapters

How children and youth live in cities and how cities are constructed in response to their presence is tightly connected to social constructions and imaginations of age and urban space. We start in Chapter 2 with a discussion of the relations between cultural constructions of childhood, youth and the city, showing the extent to which understandings of the city as an adult space rely on particular constructions of age. The chapter relates insights from the new sociology of childhood to geographical critiques of dominant constructions of the city and moves towards a relational understanding of both. While recognising the significant symbolic and material consequences of those constructions for young people, we also look towards ways of imagining the urban lives of children and youth differently, since such efforts to imagine alternative engagements with the city are crucial for creating more inclusive, participatory spaces.

The book then moves on to consider causes and forms of social inequalities that affect young people's place in the city, both in relation to adults and between different groups of children and youth. In Chapter 3, we ask why families with children are over-represented amongst the urban poor in most countries of the world, how different welfare regimes affect this uneven distribution of wealth, and to what extent class, ethnicity, gender and geographical location affect young people's well-being in the city. Finally, the diverse situations of young people living in poverty are also acknowledged, and we highlight the particular conditions of migrant, homeless and working children and youth in the city as exemplars of the distinct experiences of how inequalities are reproduced.

While the first part of the book focuses on unequal power relations that shape young people's differentiated access to and use of urban space, the second part is particularly concerned with illustrating their creative spatial practices. In Chapter 4, we focus in greater depth on this issue and discuss young people's differentiated uses of the urban environment. We start by offering some contextual background to growing up in the city through an analysis of the particular changes that occurred during periods of modernisation and urbanisation in Europe and North America. Then, taking the experiences of 'living' and 'playing' in the city, we consider the ways in which young people, differentiated by age, gender, sexuality, class and disability, engage with the urban environment in a plethora of ways that are spatial, temporal and relational.

In Chapter 5, we look in depth at the embodied, cultural practices through which children and youth contest exclusionary structures and appropriate the city, making global urban spaces at least partially their own and challenging adult domination through tactical uses of space. We examine how young people engage with transnational cultural flows, respond to the restructuring of leisure time and space, and claim a stake in urban space through subcultural activities.

Chapter 6 leads on from this to consider how urban space can be made more inclusive and responsive to young people's needs by promoting the active participation of children and youth in the planning process. We discuss the advantages and pitfalls of different modes of participation and draw on the expanding international literature on this issue to present a wide range of examples from around the world that are intended to inspire as well as to give cause for further reflection. Finally, the conclusion outlines how current and future changes, such as the emergence of the growth and shrinking of cities, can impact on young people's urban lives.

It is beyond the scope of this book to answer all of the questions that it raises. Instead, we have sought to include a number of examples and case studies from across the world that not only illustrate key arguments but can also be used as a starting point for further discussion. At the end of each chapter, we have included sample essay questions and brief lists of recommended reading, both of which we hope will be useful for engaging further with the key themes of this book.

2 Imagining children and youth in the city

In this chapter we will:

- **explore different understandings of age, childhood and youth**
- **show how these understandings relate to perceptions of rights to the city**
- **discuss how age and the city are co-constructed and which alternative representations may be constructed.**

Introduction

The design, construction and use of cities are influenced strongly by representations of people and place that reflect historically and culturally diverse understandings of age, the human body and the nature of urban space. These representations are related, as McDowell argues (1999), to constructions of citizenship and rights to the city. They create *discursive* conditions through which access to the city is negotiated, denied or claimed, a sense of belonging forged, and exclusions enforced or contested. In this chapter, we show the relevance of 'imaginary geographies' and discursive constructions of age for the design and use of cities. We focus primarily on the ways in which understandings of age influence urban design and vice versa. The relational approach advocated in the introduction also includes recognising, however, that young people can and do partake in the production of discourses about childhood, youth and the city. We give some examples of this at the end of the chapter.

One way to approach this is through the concept of *discourse* (see Box 2.1). For the French historian Michel Foucault (1972, 1980), discourses are what make the world meaningful to us. They are constructions of both power and knowledge which 'provide a language for talking about – a way of representing the knowledge about – a particular topic at a particular historical moment' (Hall 1992: 291). Discourses are not mental constructions detached from the 'real world', but part and parcel of material forms, social relations and spatial practices, which shape them and through which they manifest themselves. Through the concept of discourse, we can thus understand how 'childhood' and 'youth' emerge in particular times and places in very specific ways to produce subjects which carry meaningful identities tied to age. We can explore how the emergence of *discourses* of childhood and youth was intricately connected to institutions, scientific knowledge and state policies which produced and continue to produce subjects such as children, teenagers, adolescents or youths in particular ways and which, in turn, appear to confirm the reality of these categories by making the links between material and discursive structures concrete. Further, we can think about the extent to which our interactions as social beings are shaped by conventionalised ways of understanding and performing identities, considering which types of action are 'intelligible', or legitimate, and which seem 'unintelligible', inappropriate or illegitimate (Butler 1989, 1993).

Box 2.1 **Discourse**

The use of the term 'discourse' in linguistics, the humanities and the social sciences exceeds the conventional understanding of 'a conversation'. While it emphasises the fact that language is a social practice, 'discourse' describes how, through linguistic norms and conventions, phenomena are described, understood and even linguistically 'brought into being' in relatively regular ways in particular social settings, arenas and groups. Thus, we may talk of 'academic discourse', 'medical discourse', 'legal discourse' or 'media discourse' – arenas which have characteristic linguistic features through which their particular subject matters are conventionally discussed and made 'intelligible' to members of that group or forum, without any necessary link to an underlying, definite 'reality'. Discourses change historically and include, for authors influenced by the work of Michel Foucault, material structures and specific social relations. Discourses, like the social groups which communicate through them, do of course overlap in practice, so that certain discursive structures are shared more widely, as well as being open to contestation by 'outsiders' of the group.

In relation to childhood and youth in the city, this means not only that there are historically variable understandings of childhood and youth that connect to changing political situations and imperatives, but also that different groups speak differently (i.e. construct socially) what childhood and youth are, and where children and youth should be placed geographically (i.e. in the family home, in childcare, in educational settings ...). Medical experts might, for instance, use terms such as 'infancy', 'developmental stage, 'physique', 'paediatrics' and 'im/maturity' to communicate shared understandings of children's physical features and their own practice that enable medical research and treatment, but are interpreted differently and sometimes critiqued by social scientists, lawyers, parents or teachers, who also encounter children in very different settings. While rarely considered in debates by such adult, professionalised groups, children themselves are also part of the construction of knowledge about them and it is one of the challenges of research on childhood and youth to identify the arenas and practices through which they participate in the production of discourses about them.

The analysis of discourses of age has been central to the 'new sociology of childhood' which we discuss below. New sociologists of childhood have looked towards critical discourse analysis to unpack assumptions about essential, universally applicable characteristics of age, tracing instead how particular understandings of both age and the body have emerged historically in contingent, yet powerful ways.

Conceptualising age

Age is one of the key ways on which we categorise people and their abilities. It is a major determinant in the organisation of social groups through institutions and has a strong influence on our rights and obligations within society. When we meet a new person, one of our first questions is often about their age, because we assume that it will give us an indication of their maturity, their life-experience and how we can relate to them. It fundamentally structures our relations with people and sets up a hierarchy that is difficult to breach through individual practices. Age is shared and therefore often assumed to be simply measurable in weeks, months and years. Both in popular discourse and in much academic work on psychological and physical development, the shared and measurable characteristics of age have given rise to assumptions about 'normal' human development, which structure normative expectations of what somebody should be able to do or not at what stage in their lifecycle. For children, this often translates into finely tuned scales of skills and abilities that they should have developed at certain stages and any deviation from such scales causes concerns for experts and parents alike (Gagen 2009). Yet, even in popular discourses, understandings of age are divided by other characteristics, such as gender, ethnicity and class, which start to contradict the assumption of universal age categories. The scales along which age is measured can be helpful in identifying conditions which, with early intervention, can be alleviated, but they produce a picture of 'normal' development which more often than not classifies a range of behaviours as inferior and problematic rather than enabling us to appreciate the diversity of human life and developing ways of living with this diversity. Whether we assume that the human body will develop in a particular way independent of social conditions or that environmental circumstances hinder or help 'normal' development is irrelevant here in as far as both already assume a norm and measure individuals against it.

Relational approaches

The 'new social studies of childhood' have been at the forefront of critiquing such essentialist understandings of age (Holloway and Valentine 2000a). Arguing for a social constructionist understanding of age, its proponents emphasise the need to trace carefully how categories such as childhood and youth were invented and invested with specific meanings at particular moments in time as well as examining the subject positions that are made possible or foreclosed by changing discourses of age. Authors such as Allison James (1993; James and James 2004), Chris Jenks (2005), Christine Griffin (1993, 2001), Alan Prout (James and Prout 1997; James *et al.* 2001; Prout 2000a) and Jens Qvortrup (1994, 2005) have drawn attention to the power relations that place people unequally on the basis of age and they have been particularly critical of the denial of agency to children in contemporary Western discourses of childhood and youth. Anthropologists have further shown that understandings of age vary significantly between cultures (Amit-Talai and Wulff 1995; Christensen and James 2008).

Recent research in sociology, history, anthropology and geography has further moved towards a *relational* understanding of age, whereby age emerges from specific social relations, interactions, cultural rites, perceptions and imaginations, institutional structures, family or household constitutions, scientific knowledge and material living conditions (Alanen 2000, 2001; Prout 2000, James *et al.* 2001; Närvänen and Näsman 2004; Hopkins and Pain 2007). Närvänen and Näsman (2004) advocate the concept of the *life phase* in order to express both structural conditions and the experience of age:

> Age-related life phases, such as childhood, come into being through complex processes and are institutionalized but can also change over time (Hareven, 1995). The various life phases in a life course must be understood in relation to one another, they mutually define one another (Hareven, 1995; Imhof, 1986) … [W]hile we can discuss structural questions about the childhood life phase and its social construction, we can also use the life phase concept to study children. Children are those who are defined and/or define themselves as living in the childhood life phase. Children can then … be said to create themselves and their life phase in interaction with people who are living in other life phases … The life phase concept makes it clear that children and childhood are not only defined in relation to another category (adults), but rather in relation to other, different life phases and those who are living in them, and that the whole of this interaction can be problematized from a life-course perspective.
>
> (Närvänen and Näsman 2004: 84, 86)

Such a relational understanding of age is important if we adopt a comparative perspective that recognises the diversity of perceptions and experiences of age *between* and *within* societies across the world (Punch 2001; Jeffrey and McDowell 2004). It is further useful for analysing the relations between space, place and age. Space is constructed, lived and imagined in relation to age, while definitions of age and of a person's age-related rights and abilities entail understandings of the spaces which are deemed 'appropriate' for different age groups (Hopkins and Pain 2007; Valentine 2004). Conflicts over space thus frequently centre on the rights of different age groups to frequent and use places in particular ways. It can, however, also be a means of promoting intergenerational encounters that are of a more positive and affirmative nature (Vanderbeck 2007). Further, a relational perspective on age recognises that lives are not lived in isolation, but linked (Elder 1994; Bailey 2009), and that boundaries of age are transgressed in practice because people's relations change depending on context and requirement. Thus, not only do many children around the world work and take on caring responsibilities (Robson 2000, 2004, 2009; Evans 2005, 2011; Evans and Becker 2009; Young and Ansell 2003; van Blerk 2009), but children and adults also negotiate between their different needs and desires as they use urban space (Valentine and McKendrick 1997). At the same time, because cities are designed primarily to cater for the needs of those who drive, shop and work there, many children depend on adult carers in order to be able to, safely, access the city streets. Parents of small children, in turn, encounter barriers to fully accessing the city that rarely hamper those who are able-bodied and without children, whether it be the need for breastfeeding-friendly spaces, baby changing facilities, accessible shops and amenities for pushchairs, traffic safety, affordable transport and/or convenient car parking and, of course, a child-friendly atmosphere that makes children and those looking after them feel welcome. Child-friendly cities will thus, inevitably, also have to be cities that accommodate adults with childcare responsibilities more readily, rather than seeing them as an obstacle to the smooth running of other urban flows and processes.

Relational thinking helps us to understand age as mediated, context specific, historically variable and constructed through power and knowledge. This makes it difficult, however, to define in any essential and universal way young people's age-related rights and needs. If age categories are variable, then universal declarations of rights, such as the UN Convention on the Rights of the Child, become highly problematic (Ruddick 2007; Ansell 2005). Yet, we may want to retain a sense of age-based rights in recognition of the social conditions which continue to subordinate children and young people's rights to those of adults (Qvortrupp 1994). Calls for universal rights can be crucial for contesting the exploitation of children and young people. A relational perspective recognises, however, that activities which are deemed 'unchildlike' in some contexts are crucial for survival in others (Skelton 2007). If such activities are overlooked or criminalised, then children may be denied the very support they need the most. Indeed, much childhood labour goes unacknowledged simply because our definitions of childhood exclude the variety of labour forms in which children engage.

Defining childhood and youth

More often than not, childhood and youth are represented by adult others (Griffin 1993, 2001). Even in social research, children and young people have rarely been given much space to express their own views and to make them count. The analyst's views dominate in the representations produced for fellow academics. This has begun to change since the late 1970s and early 1980s, with the growth first of humanist research on children and youth, and later of social constructionist perspectives and the emergence of the new sociology of childhood with its emphasis on critically investigating the representation of childhood and youth, highlighting young people's agency and developing participatory approaches to research (James *et al.* 2001; Aitken 2001). Adult perspectives however, still dominate representations in the media, in political discourses and in theoretical debates, especially since established forms of representation favour literary styles of expression and institutions such as the media and academia are rarely accessible to young people.

Learning task

A simple exercise that can be done to get a sense of this is to watch the main news bulletins of a selection of TV channels and analyse them using the table below. It is best to do this over a few days and for several channels to allow for fluctuations, or to divide the task between a group of students, so that results can be compared. When completing the table below, it may become noticeable that the categories do not fit and need to be changed. Further subcategories and qualitative descriptions may also need to be added (e.g. to include more sub-issues). The 'content analysis' suggested here may lack sufficient depth and it may be helpful to answer the following questions in addition to the quantitative data collected in the table:

- Are children represented as competent or incompetent actors?
- From what angle does the camera/interviewer look at children (e.g. down/up/equal level)?
- Are children's actions represented as problematic or as positive? For whom?
- Whose concerns and interests are highlighted?
- What are the most frequent sites in which children are represented?
- What words do the journalists choose to speak to children and about them? Do these differ from adults and if so, how?

An interesting comparison could also be conducted with news programmes designed specifically for children.

	Report concerning		*Report by*		*Children interviewed*		*Adults interviewed*
	Mainly adults	*Mainly children*	*Adults*	*Children*	*With adults*	*On their own*	
Topic 1 (e.g. education)							
Topic 2 (e.g. health)							
Topic 3 (e.g. crime)							
Topic 4 …							

Authors such as Jenks (2005), James and James (2004) and Griffin (1993, 2001) have highlighted that children and youth are almost always imagined *by adults* in relation *to adults*. They are rarely thought of independently without comparison. Whether depicted in romantic or demonised terms, the starting point for analyses and depictions of childhood and youth is an assumed arrival point of adulthood. Before reaching this stage, a young person will be deemed incomplete and in need of tutelage. Being constructed as a human *becoming* rather than being (James and James 2004; Valentine 2004), the child is denied full access to social rights and seen as unable to carry adult responsibilities. In this construction, the young person is not denied agency completely, but his or her agency is seen to need channelling, putting into shape and directing along particular paths. Actions that deviate from those paths are interpreted as threatening and risky, although there are strong gender differences in the interpretation of young people's actions here. More allowance is commonly made for boys' deviant behaviour, since rebellion is seen to be part of their process of becoming a man (Mac an Ghaill 1994, 1996). Class differences and racial stereotypes also mediate the picture, as the actions of particular groups of young people are seen as particularly threatening to white, middle-class Western society.

What this shows is that, although childhood and youth are often regarded as 'natural' and universal, they are imagined not only from adult perspectives, but also through the lenses of gender, class and race. Valentine, Skelton and Butler (2003) have further shown how *heteronormativity* shapes understandings of 'normal' and 'deviant' childhood and youth. Their research gives a vivid insight into the difficult negotiations between social expectations, parental assumptions and the coming out of gay and lesbian young people. Sexuality is thus a further aspect that refracts understandings of childhood and youth. Thus, although childhood and youth appear to be shared by everyone, it is treacherous to ignore the influence of social differentiations, hierarchies and power relations on perceptions of 'normalcy' and 'deviance'. Ignoring the latter limits our ability to challenge the precise causes of particular forms of exclusion, marginalisation and oppression.

Thus far, we have examined childhood and youth as though they were identical, but there are important differences in their construction which need further examination. The categories certainly share aspects such as their definition in relation to adulthood and the assumption of a chronology that maps onto 'natural' age characteristics. They are also frequently treated as interchangeable terms, since youth sometimes incorporates childhood while 'young people' up to the age of 18 are defined as children in the UN Declaration of the Rights of the Child (1993). Academic literatures do not always help to add clarity, as young people may be classified as 'children' in the sociology and geography of childhood (Holloway and Valentine 2000), or 'youth' in much of the Cultural Studies literature (Valentine, Skelton and Chambers 1998). This indicates yet again that, as much as we may desire clear-cut definitions, age is relationally constructed and not simply measurable in year-numbers. Although childhood and youth imply successive stages in life, their boundaries are fluid and dependent on historical, social and cultural context. In this book, we follow the UN definition outlined in Box 2.3 and apply the term 'children' to young people under the age of 18, the term 'youth' to those between 15 and 24 years of age, and the term 'young people' when we speak about both children and youth. In the following sections, we seek to identify some of the specific ways in which childhood and youth are commonly defined.

Childhood

Jenks (2005) distinguishes between Dionysian and Apollonian images of childhood. These are diametrically opposed, but in everyday practice often appear in combination. The Greek god Dionysus stands for self-gratification through the pursuit of wasteful pleasures without consideration of the consequences of his behaviour. Apollo, by contrast, represents innocence and natural goodness (see Figure 2.1). The Dionysian child is seen as in need of harsh constraints and disciplining, while the Apollonian child is regarded as possessing natural goodness, needing not punishment but encouragement (Zelizer 1994). During the Victorian era in Britain, the latter image came to dominate amongst the upper classes and elite British society carved out a special place for children to treasure and protect their assumed innocence. Literature and art promoted romantic images of childhood, while at the same time new institutions, from childcare and schooling to correction facilities, brought the child into the fold of rationalisation and sought to curb its Dionysian tendencies. Child upbringing became central to nationalist projects, as the destiny of the whole nation was seen to depend on it.

Cunningham (1991) outlines the connections between the romanticisation of childhood and the emergence of urban industrialised society. He shows how the child became a projection of nostalgic desires that modern society seemed to erode, a response to the experience of upheaval and rapid social change:

> The more adults and adult society seemed bleak, urbanized and alienated, the more childhood came to be seen as properly a garden, enclosing within the safety of its walls a way of life which was in touch with nature and which preserved the rude virtues of earlier periods of the history of mankind.
>
> (Cunningham 1991: 43)

Figure 2.1 ***The angelic child.*** **Baroque statue, Potsdam, Germany. (Source: Kathrin Hörschelmann)**

Despite its depiction as natural, the Apollonian child since Victorian times has thus been produced discursively in relation to particular historical circumstances and through regulatory practices that *produce* the child-like child, not through corporeal punishment and violence but through more subtle means of correction and disciplining. Jenks (2005) refers to the ideas of Michel Foucault (1977) and Nicolas Rose (1990) to show how institutions of childhood from midwifery and post-natal care through to parenting advice, health monitoring, childcare and schooling mirror other 'panoptic' institutions, such as prisons that were designed for optimum surveillance. Through these, the Apollonian child over time learns to watch, monitor and discipline itself. Behaviour which deviates from the norms of Apollonian childhood, however, significantly undermines not only such understandings of the child-like child, but also of society as a whole:

> Any transgression in the form of childish behaviour threatens the very core of the collectivity. To be socialized is to become one with the normative social structure and so the idea of evil that is projected out into the image of the Dionysian child is, in fact providing a vehicle for expunging all sentiments that threaten the sacred cohesion of the adult world.
>
> (Jenks 2005: 79)

It is important that we remember the class-based distinctions made in the Victorian period between children as playthings of the elite and, at the other end of the spectrum, children of the poor, who were seen as a risk to and deviant from Victorian moral values. Not only did these children rarely enjoy the freedom from work that would have allowed them to idly play and explore their alleged childish instincts, but those whose parents were too poor to be able to meet even their most basic needs were excluded materially, through 'poor' schools and orphanages that were located on the fringes of the city and often resembled hospitals and prisons in their design and layout. A striking example of this is the former orphanage and school of St Peter's near Darlington in the UK (see Figure 2.2). The school was located about 15 km outside the city on a plot of land surrounded by farmland and a short distance away from a village. It was founded by the Catholic Church around 1900 and accommodated about 300 boys. The school consisted of a three-storey main building, a dormitory, gymnasium and workshops arranged around a central courtyard. This design allowed for easy supervision of the children and, because of its location, made an escape difficult. The children would have been regulated by a tight regime of work, exercise and education, designed to discipline their bodies, shape their minds and, most of all, keep them out of sight, away from any possibility of 'troubling' Victorian well-to-do society. Today, the school is still referred to as an institution for 'naughty children' by residents from nearby villages, showing that the poor and orphaned child, incarcerated in this way, became classified as delinquent and thus no longer qualified for the special protection that a 'normal' child would have enjoyed. The exclusion of such 'Dionysian' children was structural in that it allowed Victorian society to maintain romantic images of childhood undisturbed by the harsh realities of poor children's lives.

Figure 2.2 St Peter's poor school, near Darlington, UK. (Source: Kathrin Hörschelmann)

If we think only about examples from the past, however, we are at great risk of forgetting the persistence of Dionysian understandings of childhood in Western states today. A more challenging question with greater relevance to our lives today is to ask whether and how similar structures are produced currently that exclude particular groups of children (and their families) from society and thus make them exempt from the rights and opportunities that are otherwise taken for granted. Campaigns against 'antisocial behaviour', measures to moderate hyperactive children and the medicalisation of children's bodies in anti-obesity campaigns all, in different ways, testify to the persistence of images of children as risky, out-of-bounds and a potential threat to society, as well as to themselves (James and James 2004; Jenks 2005; Valentine 1996a, 2004; Aitken 2001). They subordinate children's rights and needs to those of adult society's 'moral majority' and often sideline questions of child welfare and protection in the interest of protecting 'society' from the risks that such children's bodies allegedly pose. Thus, in the UK, children as young as ten can be incarcerated and their needs as children ignored. When they transgress, they fall outside of the rules of 'normality' and enter a 'space of exception' (Agamben 2005) in which more often than not they cease to be seen as children at all. The 'deviant' child is thus excluded in order to uphold a particular image of the 'normal' child-like child (Jenks 2005). The contradiction arising from this is staggering. On the one hand, most Western countries have signed the UN Convention on the Rights of the Child. On the other hand, by exempting 'deviant' children from the category of childhood, their rights are radically undermined. In the UK, the 'ASBO kid' (see Box 2.2) can thus be prohibited from meeting friends (being denied the human right of assembly), be placed under a curfew that infringes his or her rights to access public space, and be 'named and shamed' by local newspapers, on posters and other public advertisements in direct contradiction to laws that prohibit the public revelation of a young persons' identity in case of a criminal conviction (Payne 2003). Youth offenders in custody are frequently accommodated in institutions that are unsuited to cater for their needs. In the UK, a high proportion are kept in adult prisons and in young offender institutions that are overcrowded, lack effective mechanisms to protect them from bullying and offer limited education opportunities (Goldson 1997). Not only do young offenders in such institutions suffer significant psychological harm, leading to high levels of suicide, but they are denied the education and psychological resources that would enable them to build sustainable adult lives.

Box 2.2 **Representing youth crime**

The following stories appeared on the same day in two British newspapers: the broadsheet *The Guardian* and the tabloid *Daily Mail*. They report on the court ruling against five boys convicted of manslaughter.

Learning task

To analyse the reports, note the key differences in their description of events and of the causes of the attack. What images of youth, class and gender are constructed in the reports and to what extent are stereotypical constructions of youth, class and gender made to stand in as explanations for the attacks?

Five boys convicted of killing man playing cricket with his son

Five boys who hurled stones, chunks of wood and abuse at a father playing cricket with his son, triggering a fatal heart attack, burst into tears yesterday as an Old Bailey jury convicted them of manslaughter and violent disorder. They clutched their parents in the dock at the end of a month-long trial which heard how they had carried out the vicious onslaught when none of them was over 12 and one had just turned 10. Among the youngest defendants to appear at the central criminal court, they were bailed for reports before sentencing by a judge who earlier warned about their behaviour during the hearing [...] The misbehaviour coincided with prosecution arguments to the jury that the boys' youth was no excuse for their attack and they knew their violence was wrong [...] The boys belonged to a local gang called TNE – for The New Estate – but Nicholas Valios QC, for the youngest defendant, asked the jury not to be swayed by current publicity about gang culture. 'Every day one has read about gangs killing, knifing, shooting and terrorising estates. That really isn't so in this case.' As an off-duty police officer and other passers-by tried to help Mr Norton, the court heard, the boys ran off. One was heard saying 'I think I got him', but another, in tears as he struggled to keep up with the pack, kept shouting: 'He's dead, he's dead.' Only one of the five, now 14, gave evidence, admitting that his behaviour had been 'stupid, revolting and appalling'. He told the court that he had thrown stones but only to try to topple the stumps and wreck the Nortons' game 'for a bit of fun'.

The Guardian, 1 September 2007, Martin Wainwright, p. 8

The killer aged ten

14 years after Bulger, boy convicted of stoning father to death

A boy wept in court yesterday as he was convicted of killing a man when he was only ten.

The child, now 12, was part of a gang who stoned a father of two to death as he played cricket with his son. He becomes Britain's youngest killer since Jon Venables and Robert Thompson, who were both ten years and six months when they murdered James Bulger in 1993. The 4ft-tall youngster was three months short of his 11th birthday when he and his friends launched a barrage of sticks and stones at retired draughtsman Ernest Norton, 67, in February last year [...] The boys, from Erith, Kent, were in a gang called TNE (The New Estate), who had planned a fight with rival gangs on the day of Mr Norton's death [...] Only one of the accused, now 14, gave evidence at the trial. He admitted spitting at Mr Norton and that it was 'stupid', 'revolting' and 'appalling'. But he said he was throwing stones only to try and knock over stumps and wreck the cricket game, for a 'bit of fun' [...]

Venables and Thompson, who abducted and killed James Bulger in February 1993, have now been released. In 2001 they were granted lifelong anonymity after it was argued their lives would be in danger if they were identified.

Daily Mail, 1 September 2007, Olinka Koster, pp. 1, 6–7

Feckless life of the 4ft killer

Boy had been barred from school and joined a gang

At the start of the trial, he wore the blazer of the junior school he had occasionally attended.

By its end, he was back in the hooded top, tracksuit and trainers. But any pretence of toughness was swept away by his tears. The boy who became a killer at ten has never known the man who fathered him and his older brother. His unemployed mother, now 37, married a different man a few years ago and the boys live with them in a scruffy terraced home in Slade Green near Erith, Kent.

Their mother regularly turned up at court wearing vest tops and flipflops and looking tired, with unkempt hair. Her son readily admitted to being a member of the TNE gang, but does not have any previous convictions or police cautions. The boys beside him in the dock were all born to unmarried parents. Their mothers were typically in their early 20s, though one was just 18. Those that did work had jobs in shops or restaurants. Those fathers that are known also had basic jobs – catering, building work, and tyre fitting. The boys got to know each other at school or hanging around the local estate. Graffiti has been daubed on many of the homes in the area – typically £360-a-month housing association properties. Some have signs warning: 'Danger – Dog Running Loose'. [...] In a police interview soon after the attack, the ten-year-old insisted he was happy to proceed without a solicitor, before freely admitting that he and his brother had been involved in the stoning [...] The grandfather of one of the 14-year-olds said last night he found it hard to comprehend his grandson's behaviour. [...] 'He's a loving boy. If you saw him with his baby sister you would never believe he could do something like that.'

Daily Mail, 1 September 2007, Olinka Koster, pp. 6–7

Parents

Perceptions of children *at* risk and *as* risk powerfully shape not only discourses in the media and in politics. They also have a strong hold over parents' practices of childcare, as Valentine's extensive research on this subject has shown. Concerns about children's safety in public space have increased substantially over the past few decades, partly because of increasing traffic and partly because of high-profile cases of child-abduction. Parents' fear for their children's safety impacts on the spatial range of children's activities, the organisation of their daily lives, their access and use of public and private spaces, the type of play they can engage in, etc. It also impacts heavily on parents' lives as supervising children's activities takes up more time in the day. Valentine (1997a and b) describes the social pressure parents are frequently under to adopt particular parenting styles:

> The state, medical care professions, the media, and peers (to name but a few) all play a part in dictating what is good for children and how they should be looked after; in other words, what it means to be a 'good parent'. Indeed, fathers, and in particular, mothers, are commonly blamed for their offsprings' accidents, misbehaviour, crimes and so on.
>
> Popular discourses about appropriate ways to raise children are produced around many different issues – from what they should be fed or clothed in, to what they should be allowed to watch on television [...] Parents determine the extent of their children's personal geographies by deciding at what age they should be allowed outside alone and at what age, and when, they may go to different places (the shops, school, the park, the city centre, etc.) unaccompanied by an adult. What it means to be a 'good parent', therefore, is to walk a tightrope between protecting children from public dangers by restricting their independence, whilst simultaneously allowing them the freedom and autonomy to develop streetwise skills and to become competent at negotiating public space alone. It is a process that is highly gendered, both in terms of parents' perceptions of boys' and girls' respective vulnerabilities and competencies to handle danger in public space; and in relation to how mothers and fathers negotiate the parental responsibilities of setting children's spatial boundaries and discipline any infringements.
>
> (Valentine 1997b: 37–38)

Parents are quickly castigated as irresponsible and uncaring if they leave their children unsupervised for lengthy periods of time. 'Teenage mothers' and 'single parents' are particularly typecast as 'bad parents' for the very reason that constraints on their time and budget may not allow them to provide supervision for their children around the clock. Moral anxieties to protect children (and to protect society *from* children) conflict, however, with perceived developmental needs of children, such as the need to learn about the environment by exploring it on their own, by taking certain risks in order to learn about and improve their own capabilities, to interact with other children and to enjoy a certain amount of freedom to roam. Further, perceptions of children's needs and capabilities, as well as parenting cultures, are mediated by class, gender and ethnicity, as well as by the changing constitution of households.

Perceptions of children and young people *as risk* also have a significant impact on parents' fears (Valentine 1997a and b; Valentine and McKendrick 1997):

> Moral panics about everything from child murderers to teenage gangs, to joy riding and juvenile crime rates, have been used to fuel adult fears that public space is being overrun by violent and unruly teenagers who are a threat to the personal safety of young children, women and the elderly and who are disrupting the moral order of the street (Valentine 1996a, b). Thus, a significant aspect of the contemporary debate about children's play centres on the issue of parents' ability to control their offspring in public space. On the one hand parents are concerned about children's (particularly those under 12) vulnerability to violence (from older children and adults) and therefore seek to control their play in order to minimise their exposure to danger. On the other hand parents are anxious that older children may become embroiled in smoking, drug taking, drinking or underage sex, and therefore seek to control their teenagers' outdoor leisure activities in order to minimise the risk that their offspring will harm themselves or others.
>
> (Valentine and McKendrick 1997: 223)

Parenting styles are thus not only impacted by fears about dangers to children, but also by anxieties over the changing 'nature of childhood' (Valentine 1996a and b) in relation to both risk-taking and risk-posing behaviour. They thus meander between both Dionysian and Apollonian perceptions of childhood.

Youth

The 'deviant' child is frequently excluded from childhood not only through institutional structures, policies and the design of cities, but also temporally, by being reclassified as an adolescent or teenager, a child that is growing up but has not yet reached the maturity of an adult. Adolescents are said to occupy a transitional stage between childhood and adulthood during which they need to achieve a number of developmental tasks in order to leave childhood behind forever and to reach maturity. This transitional stage has been described by psychologists in the tradition of Stanley G. Hall (1904, 1906) as both necessary and confusing for the young person who may be drawn between a desire to cling onto the carelessness of childhood whilst wishing to already exercise the same rights as adults (Erikson 1968; see Epstein 1998). Like childhood, youth is however a fairly recent invention that has as much to do with the growth of education and the discovery of young people as consumers as it has with physical and psychological development (Valentine *et al.* 1998). In Western societies, it most commonly describes children from the age of secondary school onwards and terminates (with much variability) around the time when the young person gains employment, has children of his or her own, becomes financially independent, enters sexual relationships, etc., although a 'teenage parent' may be castigated as irresponsible, too young to have children and a burden for society, thus not accepted into the adult world. In many southern African societies youth policies by contrast include young people up to 28 or 30 (occasionally 35), while culturally marriage and childbearing at any age results in adulthood. The UN defines youth as the period between 15 and 24 years of age, but includes young people under the age of 18 in the category of 'childhood' in its Convention on the Rights of the Child (see Box 2.3).

Box 2.3 **UN definition of youth and childhood**

The United Nations, for statistical purposes, defines 'youth', as those persons between the ages of 15 and 24 years, without prejudice to other definitions by Member States. This definition was made during preparations for the International Youth Year (1985), and endorsed by the General Assembly (see A/36/215 and resolution 36/28, 1981). All United Nations statistics on youth are based on this definition, as illustrated by the annual yearbooks of statistics published by the United Nations system on demography, education, employment and health.

By that definition, therefore, children are those persons under the age of 14. It is, however, worth noting that Article 1 of the United Nations Convention on the Rights of the Child defines 'children' as persons up to the age of 18. This was intentional, as it was hoped that the Convention would provide protection and rights to as large an age-group as possible and because there was no similar United Nations Convention on the Rights of Youth.

Source: UN (no date) *Youth and the United Nations*, http://www.un.org/esa/socdev/unyin/qanda.htm, accessed 16/07/08

Griffin (1993), echoing the work of Cohen (1972), Willis (1977) and other theorists from the Birmingham Centre of Contemporary Cultural Studies (BCCCS), has argued that the category 'youth' was invented and gained currency in response to societal crisis in nineteenth-century Europe and North America. It expressed anxieties about rebellious working-class young people, who were constructed as an 'Other' to middle-class adult society and invested with meanings of danger and risk that justified the extension of institutional means to discipline and 'correct' unruly youths. The treatment and management of youth was seen as essential for the maintenance of 'the nation' and thus became a central societal task. Griffin identifies a range of discourses which construct youth even today: a clinical discourse, a discourse of criminality, a discourse of education and a discourse of sexual deviance and disaffection. Despite its variability and relatively recent invention, '[y]outh/adolescence remains a powerful cultural and ideological category through which adult society constructs a specific age stage as simultaneously strange and familiar. Youth/adolescence remains the focus of adult fears and pity, of voyeurism and longing' (Griffin 1993: 22).

Envy and fear are expressed today in a complex mix of promoting the virtues and possibilities of eternal youth on the one hand, while producing anxieties about under-age drinking and sex, gang culture and violence, or anti-social behaviour of young people on the other hand. Yet, as Giroux critically comments,

> Youth as a complex, shifting, and contradictory category is rarely narrated in the dominant public sphere through the diverse voices of the young. Prohibited from speaking as moral and political agents, youth become an empty category inhabited by the desires, fantasies, and interests of the adult world ... Lauded as a symbol of hope for the future while scorned as a threat to the existing social order, youth have become objects of ambivalence caught between contradictory discourses and spaces of transition.
>
> (Giroux 1995: 24)

Their designation as deviant 'Others' means that the actions of young people are quickly defined as rebellious and/or dangerous. Yet, their marginalisation also means that young people may indeed engage in more radical actions to challenge their own exclusion and/or that of others.

The notion of 'radical youth' is undermined, however, by the category of the youthful consumer (Coleman, 1961). The 1950s saw the emergence of this new category with goods and services aimed at apparently fun-loving 'teenagers', whose distinctiveness was underlined and redefined in commodity culture rather than erased (Valentine *et al.* 1998). Marketing aimed at young consumers has repeatedly picked up rebellious, underground trends in a search for novel ideas and products. By mainstreaming these for a broader audience, it undermines at least in part their radical potential (see Chapter 5).

Implications for imagining children, youth and the city

Thus far, we have considered understandings of childhood and youth in relative isolation from spatial conceptions. Yet, there are close connections between the representation of spaces and imaginations of identities, including those of age. Thus, not only do many institutions (schools, retirement homes) relate to specific age groups, but access to spaces such as the street, shop, pub, club, playground or park is regulated, and contested, on the basis of age. Imaginations of wider landscapes such as the city and the countryside also relate to constructions of age-based identities, in as far as certain landscapes are seen to be more or less appropriate for particular age groups and those landscapes in turn might be described in age-related terms (e.g. the young and vibrant city; traditional country life; the dying city/village, the suburban family home). Understandings of age-characteristics map onto representations of spaces and vice versa, though in no unitary and unambiguous way. Representations of age and space are intricately connected to questions of politics and the negotiation of rights to/within/over space, since exclusion from and inclusion in particular spaces will often be related to definitions of age.

Understandings of urban childhood and youth intersect with and in many ways reflect imaginations of the city. While for some, cities are intrinsically progressive, the traditional view of cities and their relationship with people has been negative (Knox and Pinch 2006; Short 1991). As Knox and Pinch note,

> Public opinion and social theories about city life, together with the interpretations of many artists, writers, filmmakers and musicians, tend to err towards negative impressions. They tend to be highly deterministic, emphasizing the ills of city life and blaming them on the inherent attributes of urban environments …
>
> (Knox and Pinch 2006: 152)

Representations of the rural and the urban tend to fall on one side or other of a set of polarities (Knox and Pinch 2006; Fisher 1976; Short 1991), such as:

Rural		*Urban*
Nature	versus	Culture/Art
Familiarity	versus	Strangeness
Community	versus	Individualism
Tradition	versus	Change

(Knox and Pinch 2006: 152)

Neither side is intrinsically 'good' or 'bad'. Such interpretations depend strongly on one's political and personal beliefs. However, when related to childhood and youth, we quickly see how they map onto binary images of the Dionysian and Apollonian child. Characteristics associated with the city rarely reflect idealised notions of childhood and they intersect with constructions of youth in predominantly problematic ways, though transitions to adulthood are frequently seen as necessitating at least a temporary move to the city (Beck and Beck-Gernsheim 2002).

Cities are rarely regarded as the best places to bring up children (Valentine 1997a and b; Jones 1997, 2002). Although they may be seen to hold opportunities for encounter, independence and change for adults, for children and young people they are more regularly regarded as threatening places, where stranger-danger looms large and where the density, scale and intensity of urban flows and networks disorient and pervert the innocence of children. Since the latter are assumed to be particularly close to nature and to need access to nature for 'normal' development, cities are classed as unsuitable environments for them. Paradoxically, such imaginations ignore the fact that the vast majority of children and young people do, of course, grow up in cities today. The 'rural idyll' has not only been shown to be an ideological construction at some remove from the realities of rural life, but it simply has no relevance for understanding the actual lives of urban children and young people. Yet, it has a strong hold on adult reveries and imaginations of an ideal childhood (Philo 2003; Jones 1997).

Fear of the city, children at risk and risky youth

Amongst the negative aspects of cities that are thought to be particularly threatening to children and young people are traffic dangers, pollution, lack of community interaction and the surveillance that comes with it, dangerous strangers, obstacles to disciplining and delimiting their activities in urban space, getting lost and being without orientation. Parenting cultures, as we have seen above, reflect this perception of the dangers of urban space and have become more restrictive in Western societies over the last few decades. Fear for the safety of children and young people motivates night-time curfews, restrictions on their independent and unaccompanied movement, and the growing privatisation and enclosure of play-spaces at a time when public facilities are increasingly replaced by commercial leisure spaces that provide sanitised, supervised spaces for play removed from the apparent dangers of the street (Valentine and McKendrick 1997; McKendrick *et al.* 2000a and b, see Chapter 5). Over-protection of children and young people from perceived threats of the city, however, can have the effect of reducing their very ability to act competently in the city (Katz 2006). It also distracts from tackling the causes of some of those threats:

> [T]he 'modern passion for safety', which resonates with and against the terror talk about young people, disciplines children, literally by keeping them indoors when unsupervised, and figuratively by keeping them surveilled when outside. It disciplines women by suggesting – slyly or otherwise – that unattended children are at risk; the sub- (or not so 'sub') text of worrying narratives about children's vulnerability blames mothers for leaving their children in the care of others … The unease with a nonwhite planet, with women working outside the home (unless they are welfare recipients in which case their mothering activities are constituted as lazy moochery), and with the confusing manifestations of 'globalization' come home, have produced a fearsome, vengeful, and increasingly privileged ruling class … One of its faces is witnessed in the abandonment, criminalization, and incarceration of large segments of the youthful population, particularly young men of colour, with little promise of meaningful work in the quickly restructuring 'global' economy … But there is another face to this process. One I've suggested can be read in the landscape. That is privatization, which is effectively reordering the nature and spaces of social reproduction.
>
> (Katz 2006: 116–117)

As Katz suggests above, children and particularly adolescents from underprivileged backgrounds are also seen as a risk *to* the city. Their behaviour in public space is forever monitored and regarded as potentially or actually disruptive. Even the sheer presence of groups of young people on central squares, in parks, shopping centres and on the street is often perceived as threatening and the view that children should be seen but not heard assumes its full meaning in adult complaints about noisy youths populating urban space late at night. The case study of antisocial behaviour measures in the UK (see Box 2.4) shows clearly how 'yobbish youngsters' and 'feral children' are constructed as one of the major threats in the city. Such campaigns demonise all children and young people as potentially deviant and risky. They blur the distinction between fear of crime and actual crime, broaden the spectrum of what is seen as deviant, punishable behaviour, and establish a strong distinction between law-making adults and law-receiving minors, where the latter have no voice but are made to carry the full weight of adult decision-making. They are excluded on the basis of age and prejudged through general stereotypes based on *fear* rather than actual crime. It is telling that the 'Respect' campaign was launched by the former UK Prime Minister Blair in 2006 with the expressed aim of reducing *fear* rather than crime (Respect Task Force 2006; see Box 2.4).

Figure 2.3 Sign outside main station, Potsdam, Germany. (Source: Kathrin Hörschelmann)

Adult fear *of* rather than *for* children and young people provides the justification for their exclusion from all manner of public spaces, while private venues have also introduced new surveillance and exclusion measures to control young consumers who are clothed and/or act 'suspiciously' (Kato 2009; Alexander 2009). The fact that such surveillance techniques target children and young people for who they *are* rather than for what they *do* as individual actors is not only discriminatory in the extreme. It also shows that it is young people's right to public space *per se* which is at issue. Policies to survey young people's actions in urban space and to punish even minor transgressions severely are tied to the emergence of what Neil Smith (1996) has called 'revanchism', a populist right-wing backlash against liberalism and the politics of state welfare which mixes military-style tactics with moral discourses about public order and extends the policing and privatisation of space into areas previously thought of as 'public' (see Figure 2.3). As Mitchell (2003) and Katz (2004, 2006) have argued, revanchist policies have major implications for social justice, understandings of 'the public' and citizenship. As the diversity of users of public space is reduced, social conflicts that would otherwise have erupted and been negotiated within shared spaces are edited out, moved on and dealt with by exclusion rather than engagement (Malone 2002). The reasons given for increased policing, surveillance and privatisation sound appealing in the context of terror-talk and the 'passion for safety'. They promise trouble-free enjoyment of the city without the need to confront society's ills and conflicts. Yet, we are rarely asked to consider exactly *who* and *whose interests* have been edited out, or *why*. And, of course, the rhetoric of terror and risk is built on the promise that the excluded will always be Other, not the 'law-abiding' self.

Box 2.4 **Anti-Social Behaviour Orders (ASBOs) and the discourse of youth as risk in the UK context**

The juvenile justice system has changed dramatically in the UK over the last two decades, with increasing rates of incarceration for young offenders and the introduction of new, community based measures such as ASBOs (antisocial behaviour orders). As early as 1997, Barry Goldson (1997) argued that there had been a major shift from a concern with justice to one of punishment and condemnation. The U-Turn was partly fuelled by a new discourse of liberalism having gone 'too far' and by the moral outrage that followed the murder of three-year-old James Bulger, who was murdered by two ten-year-old boys in 1993. New custodial sentences were introduced for ten- and 11-year-olds with the 1998 Crime and Disorder Bill (Hodgkin 1998; James and James 2004), and a night-time curfew for under-tens was implemented. Hodgkin (1998: 67) critiqued this aspect of the Crime and Disorder Bill sharply, pointing out that civil liberties are tied up with rights to public space:

> First it can be argued that the curfew is a breach of the European Convention on Human Rights. This provides that no-one (including children) can be deprived of their liberty without proper orders on specified grounds, that the state cannot interfere with parents' rights to bring up children as they see fit (compatible with human rights) and that everyone has a right to freedom of association with others. [...] Second, the curfew is an affront to the child protection principles of the Children Act. A blanket curfew sweeps all under-10-year olds under the carpet. Of these, some may be staying out at night for very good reasons, such as a drunken and violent parent or a sexually abusive one. [...] Third, the curfew is a punitive measure which will not bring children the support that they need and is likely to increase hostility towards them.

The 1998 Bill introduced a whole suite of new punitive measures, including Anti-Social Behaviour Orders (ASBOs), aimed at tackling 'behaviour that causes or is likely to cause harassment, alarm or distress to one or more people who are not in the same household as the perpetrator' (Home Office 2006). By 2006, 7,356 ASBOs had been issued, 55 per cent to adults and 43 per cent to juveniles (BBC News 2006). The definition of antisocial behaviour is wide and encompasses, amongst other things, graffiti, abusive and intimidating language, excessive noise, litter, drunken behaviour in the streets and resulting mess, dealing drugs, etc. (ibid.). Worryingly, while most victims of violent crime have been attacked by someone known to them and the most serious crimes against children and young people are frequently committed *within* households, antisocial behaviour policies focus public attention on issues that are arguably much less serious and that can be tackled through existing legislation, if they result in grave consequences (such as drunken violence) (Payne 2003). Antisocial behaviour policies water down definitions of crime and give greater powers to authorities outside the criminal courts.

In the Anti-Social Behaviour Bill of 2003 police were given powers to disperse groups of two or more people and to impose a night-time curfew on unsupervised young people under 16, removing them from public space to their place of residence between 9pm and 6am (Payne 2003). Antisocial behaviour measures thus directly intervene in the regulation of access to public space and explicitly seek a solution to low-level crime in a spatial politics of exclusion and of confining young people to 'private' space, irrespective of the conditions of that space. It is very much a policy of 'out-of-sight, out-of-mind' that has been severely critiqued for not tackling the underlying causes of crime, for demonising children and young people, for penalising them as a group rather than on an individual basis, for breaching both the Human Rights Act 1998 and the United Nations Convention on the Rights of the Child (e.g. by curbing the freedom of association, Payne 2003), and for diverting attention from issues that continue to impinge on children's and young people's welfare, such as child poverty, violence, abuse and high rates of incarceration.

Alternative representations

Fiction for children and youth

We should not assume that all representations follow simple binary divides. Examples of the many ways in which images of childhood and youth in the city are either mixed or transgressed can often be found in representations that are aimed at a young audience. In children's fiction, for instance, we find a wide range of characters and often an emphasis on children's resourcefulness, their ability to tackle the challenges of urban life with wit and creativity and a carving out of independent spaces where adult rules are transgressed because they are arbitrary, outdated or oppressive (see Box 2.5). While we focus on fiction here, you may also want to think about films or theatre plays you have seen that develop different imaginaries of young people's lives in the city.

On closer inspection, it is noticeable that many of those stories assume a male key protagonist and mirror gender stereotypes but there are examples, such as Roald Dahl's *Matilda* and Astrid Lindgren's *Pippi Longstocking*, where girls occupy centre-stage and/or act in less stereotypical ways. In examining fiction for children we must not conclude, however, that the stories represent childhood from the perspective of a child's imagination. Novels written by adults for children say more about their own understandings of childhood than about children themselves. Philo (2003), Jones (1997, 2002) and Bavidge (2006) have argued this point strongly. Thus, for Philo, the place to look for the 'stuff' of childhood are not adult-written stories but the tales children tell themselves (2003). Adult fiction for children mediates particular ideologies of childhood and about the world to children. It thus needs to be seen as an interested, not an impartial construction.

The 'stuff' of children's literature is made up not only of characters but also of the places through which their actions evolve. Those places are not simply a backdrop or stage but in many ways inform how we imagine the central figures and what they can do. When we look closely at fiction for children, however, we find that the myth of children's closeness to nature is often reproduced by locating these stories in rural spaces that seem more natural and therefore closer to children's own nature. Jones (1997, 2002) and Bavidge (2006) have noted that children are rarely shown to inhabit urban spaces in creative and life-affirming ways:

> Children are largely excluded from accounts of the city, either literary or theoretical, and when they are present their roles are strongly circumscribed, especially given the powerful association of childhood with the rural and natural [...] [T]he child and the city are commonly seen as incompatible entities.
> (Bavidge 2006: 323)

Nonetheless, analysing fiction for children is useful for showing 'the discourses by which places are made visible in children's literature, the uses that are made of the figure of the child in writings about urban worlds, for example, or pastoral stories, and the way the narrative logics and representative strategies of children's literature have their own spatial politics' (Bavidge 2006: 323). Jenny Bavidge further argues that in bringing together the child and the city, some novels offer models for children's urban agency and, by drawing on the discursive 'otherness' of children, render the city 'other' too (also see Jones 2002). She cites the example of Louis Fitzhugh's *Harriet the Spy* (1964):

> Harriet's adventures take her well beyond the constricted space of her own home. She breaks and enters houses and takes rides on dumbwaiters, sneaks through back allies and peeps into windows. She barely avoids getting caught. Harriet's adventures occur in public space (not the private space of the secret garden), a populated environment [...] Yet, her adventures are not so much direct struggles with opposing forces (as might be found in a boy's adventure) as covert operations to ferret out knowledge of social relations.
>
> (Bavidge 2006: 328)

Box 2.5 **Detective Pinky**

Pinky is an orphan living in Kittsburgh who dreams of becoming a famous detective. He knows the city inside-out and upside-down, moving through urban space through back alleys and over the roof tops, jumping, crouching, sliding, ducking. With wit, ingenuity and a sense of justice, he solves criminal cases and helps his friends to improve their difficult lives in the orphanage, always in contest with adults who try to discipline the children.

> Pinky sat on his rubbish bin and dreamt. The bin was painted in crazy colours and stood on the roof of the house. Pinky had found it. Well, not really found it, but also not really stolen it. One morning, it was just lying in the middle of the street. He moved it to the sidewalk at first. Not like he was particularly tidy, but he liked cars and the thought that one might come round the bend too fast and crash with the rubbish bin worried him. When the rubbish bin was still on the sidewalk that afternoon, he rolled it into the courtyard and placed it in-between the clutter that was lying everywhere. He waited for two days and then lugged it up to the roof, scrubbed it clean, painted it and declared it to be his throne.
>
> So there he sat, six floors above the city, leaning against a chimney, letting the sun shine into his face and dreaming his favourite dream: the whole city of Kittsburgh had gathered in Central Park, a band was playing, the mayor raised his hand, a groan went through the crowd, the cover slipped and revealed the monument. And there was Pinky in marble on a pedestal on which was written: 'From the city of Kittsburgh to her great son'.

That evening, Pinky and his friend, Monster, are punished for returning late to the orphanage and the caretaker, Potter, charges them with cleaning the attic and collecting bundles of clothes from the community hall. They are trying to get to the circus however, where Cindy, the daughter of an acrobat has promised to let them in for free through the back-door, if they are on time.

'What a mean bastard', Monster cried as soon as Potter had closed the door behind them. 'Good bye, Cindy, good bye, circus! We need at least quarter of an hour to get to the community hall, down the whole avenue and back on 17th street, and with those bundles it will take even longer!'

'Let's take the flight route', Pinky said and ran up the stairs. 'We can't let a few worn trousers get in the way of seeing the circus. It's bad enough that we need to run around in second-hand clothes.'

'Well, we're just second-hand guys', Monster swore.

They had often run across the rooftops and knew every inch up here. But it was a different matter to be jumping around with big bundles of clothes on their backs, along the narrow edges and around the chimneys, six floors above the ground, and then they had to jump across a gap of five feet between two houses.

'I'll go first', Monster said, 'and you'll throw the bundles to me, but watch out that they don't land in the courtyard.'

Source: G. Prokop (1984: 5–10), adapted and translated by K. Hörschelmann

Representations by young people

While young people are rarely given sufficient access to media outlets and publishers to produce accounts of their urban lives from their own perspective, increasingly urban planners, artists and researchers look for creative ways of engaging young people in dialogues on urban design that have the potential of challenging understandings of childhood and youth too. We return to this issue in more depth in Chapter 6, but outline in Table 2.1 some of the methods that can be used to give children and youth of different ages, with different abilities and interest 'voice' in urban planning projects and research. Box 2.6 further demonstrates some of the difficulties that are involved in analysing visual representations, even though these are often seen as more accessible or self-explanatory.

Table 2.1 Methods for research with children and youth

Method	*Description*	*Key benefits*	*Challenges*	*Suggested reading*
Photograph diaries	Young people use cameras to take a series of pictures representing different aspects of their lives in the city.	Gives young people control in investigating urban environments. It is generally thought of as a 'fun' method that young people are enthusiastic about.	Taking pictures can raise suspicion among passers-by and can cause friction if people are included in the pictures. Some pictures are poor quality or it can be difficult to see what they represent.	• Dodman (2003) • Young and Barrett (2001)
Mental maps	Young people are asked to draw a map of their use of the city or their spatial awareness of a neighbourhood.	Can be used to find out about spatial range and young people's socio-spatial connections to urban places.	Depending on the drawing abilities of the young people, this method can be a difficult task.	• Potter and Wilson (1991) • Young and Barrett (2001)
Child-led interviews	Children interview each other using a pre-prepared schedule.	Can be useful for gaining detailed information on specific aspects of urban life. Usually provides rich in-depth data.	Can be difficult to keep on track. How do you ensure the interviewer follows up questions?	• Hecht (1998) • Young and Barrett (2001)

Focus groups	This can be child-led. Young people sit in a circle to discuss a set of issues related to the topic. Children can take turns at facilitating parts of the discussion under the direction of the researcher.	Discussion with peers can help to overcome problems of non-responsiveness and a group setting can encourage people to speak. Young people often cross check what each other are saying.	Can be dominated by particular individuals. How might you reduce these unequal power relations in the group?	• Barker and Weller (2003)
Diaries	Young people keep a log in words and pictures of their daily activities.	Very detailed information can be recorded without constant presence of a researcher. Thoughts and feelings can also be noted.	Can be seen like school or homework. How do you overcome this authoritarian approach?	• Barker and Weller (2003)
Drawings	Young people are asked to draw a picture or series of pictures representing an issue in their lives.	This is a good method for young children, whose verbal and/or written expression may be limited.	It is not always easy to see what has been drawn and drawings have different meanings in cross-cultural settings.	• Punch (2001) • Young and Barrett (2001)
Questionnaire	A series of generally closed questions about specific aspects of a topic.	Can generate large quantities of data useful for supporting more qualitative research.	Young people might not always understand the questions being asked.	• Boyden and Ennew (1997)

Box 2.6 **Analysing visual methods**

Numerous researchers that have used visual methods in their research with children and youth advocate the benefits of these methods as 'fun', less obtrusive into young people's lives and useful for overcoming language barriers when working cross-culturally (see, for example, Punch 2002; Young and Barrett 2001). However, analysing these images can be far from straightforward. Samantha Punch (2002) notes that often images that seem perfectly reasonable in the field can become distorted on return to the university and what seemed easily detectable as a house becomes less clear. Houses look like schools and modes of transport begin to look the same. It is clear then, when conducting research with visual methods that it is important to involve the artist in the analysis of the image, finding out what was drawn and why. Often images can actually represent something totally different to what appears on the paper and for this reason oral description is both necessary and illuminating. The following two examples demonstrate this well.

Figure 2.4 Example 1: street life in Kampala, Uganda. (Source: Lorraine van Blerk)

This picture of downtown Kampala was taken by Peter, a 12-year-old street child, to show an important aspect of his life in the city. It is not easy to see what he was photographing but suggestions could include the centrality of rubbish skips to their existence. However, Peter actually photographed another child pick-pocketing from a passer-by and discussed this in his oral description.

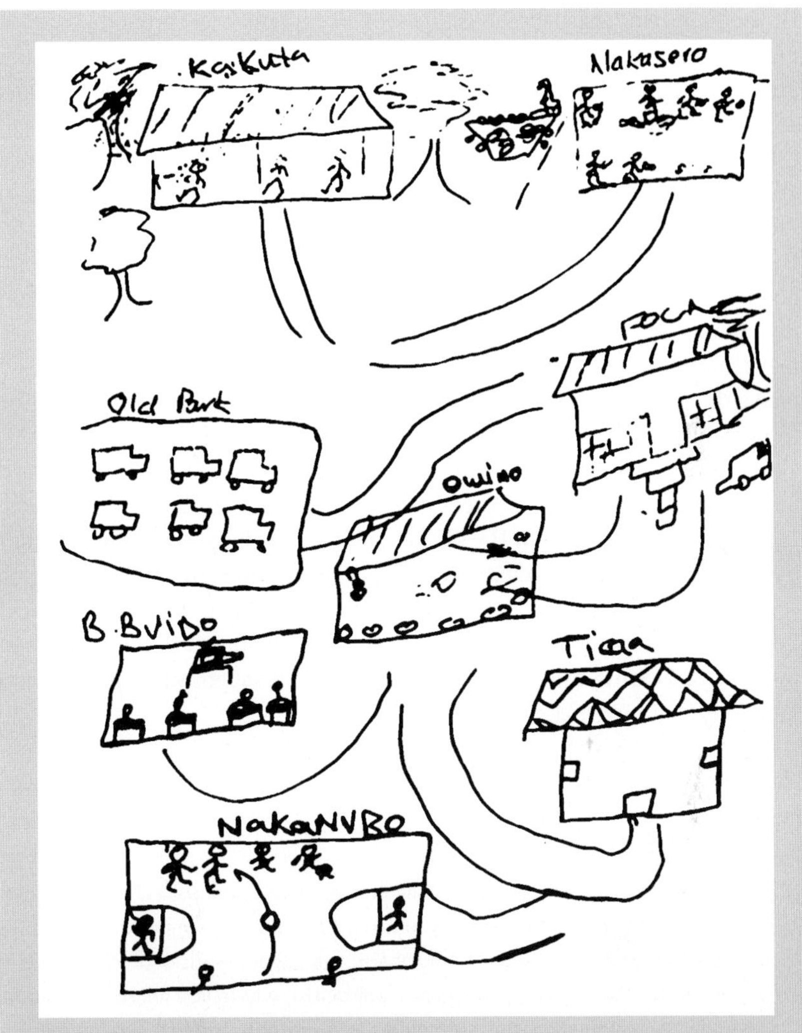

Figure 2.5 Example 2: Bob's (15) map of street life in Kampala. (Source: Lorraine van Blerk)

Bob's map highlights all the important places where he spends his day in Kampala as a street child. Although we can clearly see that he frequents the football stadium (Nakanvbo) or the video hall (BB vido), what we do not get from the images is why these places are important. For example, through asking about the map we are able to learn that Bob goes to the video hall in the evenings as this is the time when the police are around and he is more likely to be arrested. From including Bob in the analysis the richness of the data is enhanced.

Beyond applying these methods that seek to explicitly foreground young people's perspectives, it is important to recognise, however, that young people inevitably already participate in discourses with parents, teachers and other adults that evolve around their (contested) uses of the city in everyday life (Valentine 2004). Kallio (2007, 2008), for instance, explores the politics of very young children which, she argues, is expressed in the performance of their embodied identities and in their more 'mundane', yet nonetheless political, uses of urban space. Van der Burgt further reminds us that children also construct their own images of urban neighbourhoods, which include perceptions of some areas as more dangerous than others. Based on her research in a mid-sized Swedish town, she concludes that children 'construct representations of their own neighbourhoods as "quiet" places in relation both to their personal knowledge of the neighbourhood and to perceptions of "non-quiet" people and places somewhere else' (2008: 257). Children are aware, and respond to, media discourses and representations of their neighbourhoods by others. Places that are frequently stigmatised as dangerous or 'non-quiet' like the area of Backhagen, in van der Burgt's example, are defended by the children who live there, as well as subdivided further: 'While the children resist dominant discourse by placing themselves in the quieter part of Backhagen, they also use the dominant discourse on other inhabitants of the neighbourhood by placing them in the not-so-quiet part of Backhagen, turning resistant power into dominant power' (2008: 267).

Learning task

Draw a mental map of places in a city that you knew well as a child and annotate the map to show why the places you have included were important for you, how you remember feeling when being there, what you used to do there and with whom, what made them attractive or unattractive, where you felt safe or in danger and why, how you interacted with others (e.g. parents, friends) in different places. Now consider what might have influenced your recollections and how the recollections of others might differ from yours. What other methods could you have used to represent your remembered places and what are the advantages and drawbacks of different methods for representing children's spatial practices and important places? What things, feelings, memories are difficult to represent at all?

Horton and Kraftl (2006) finally argue that focusing primarily on representational aspects of the geographies of childhood and youth leads to a narrow conception of their diverse identities and uses of space. The authors argue for greater engagement with theories of embodiment, affect and performativity in order to develop a 'deeper, more detailed or expanded apprehension of the varied geographies of children and young people' (ibid.: 75; also see Colls and Hörschelmann 2009; Hörschelmann and Colls 2009). Paying greater attention to young people's embodied spatial practices and to their seemingly mundane, everyday lives not only enables more diverse stories to be told, but also to identify and challenge practices of marginalisation and exclusion that operate through, and at the level of the body, but are frequently hidden from view because they are difficult to articulate and validate. The experiences of young people with disabilities are a case in point (Holt 2004; Skelton and Valentine 2003), as are the politics of the body involved in gendered uses of space and in the performance of different sexualities (Valentine *et al.* 2003; see Chapter 4). In order to develop a relational understanding of childhood, youth and the city that recognises the diversity of young people's urban lives and their participation in the production of cities, we do indeed need to pay greater attention to the nuances of their everyday geographies. This includes searching for new ways of articulating them (Horton and Kraftl 2006; Latham 2003).

Conclusion

This chapter has explored how conceptualisations of childhood, youth and the city come to influence the lived experiences of young people in the city. By examining the different ways childhood, youth and the city are thought about both conceptually and practically, the chapter has demonstrated that it is unsatisfactory to think about age as categorised simply by biological characteristics and developmental stages. Reading childhood and youth historically and culturally social scientists now argue for a relational approach to age through which childhood and youth are perceived to be socially constructed categories based on specific social relations, and interactions of culture, politics and institutional structures at any particular time.

This means that in exploring the urban lives of children and youth we need to critically examine the ways in which children and youth are represented by others and how such representations intersect with images of the city, in different cultural contexts. Young people do, however, also produce their own accounts of the city and a relational perspective means not only recognising this but showing the extent to which children and youth, through their own imaginations and participation in discourse, partake in the making of urban space. It further means acknowledging that discourses neither occur in a vacuum nor capture all there is to the geographies of children and youth. In the following chapters, we show

how discourses of childhood, youth and the city intersect with wider social and spatial structures that set the material context in which young people grow up in diverse cities around the world, while not completely foreclosing opportunities for young people to appropriate urban space and to make the city, partially, their own through creative spatial and cultural practices.

Questions for discussion

1. What has been the effect of the social constructionist critique of chronological approaches to age on understandings of children's identities and their place in the city?
2. To what extent is young people's exclusion from public space a result of discourses of youth *as* risk?
3. Why is the city often imagined as a space *of* risk for young people and on what grounds can this be challenged?

Suggested reading

Aitken, S. (2001) *Geographies of Young People: The Morally Contested Spaces of Identity*, London: Routledge.
Amit-Talai, V. and Wulff, H. (1995) *Youth Cultures: A Cross-Cultural Perspective*, London: Routledge.
Ansell, N. (2005) *Children, Youth and Development*, London: Routledge.
James, A., Jenks, C. and Prout, A. (2001) *Theorizing Childhood*, Cambridge: Polity.
Jenks, C. (2005) *Childhood*, London: Routledge.

Useful websites

Geographies of children, youth and families: http://www.gcyf.org.uk/
Childwatch International research network: http://www.childwatch.uio.no/
Child Rights Information Network: http://crin.org/

3 The causes and effects of social inequalities on children and youth in the city

In this chapter we will:

- **explore the causes of social inequalities on children and youth in the city**
- **discuss the effects of living in poverty for young people's lives**
- **examine the outcomes of poverty on young people's lives through their experiences of work movement and homelessness.**

Introduction

The twenty-first century has been hailed as 'the century of the city' with half the world's population already living in cities and the number is set to rise to 60 per cent within the next two decades (UN-HABITAT 2008). As we note throughout this book, cities are dynamic places, ever changing, responding to global processes and striving towards the future. However, the consequence of this never-ending journey is the contradiction between cities as drivers of an economy's wealth and social development, and as a repository of poverty, exclusion and environmental problems. It is in urban areas that stark contrasts between rich and poor are most prevalent and, with the fastest urban growth rates in cities of the Global South, addressing social inequalities in relation to the physical, social/cultural and environmental fabric of people's lives is a key issue. It is not surprising then that the UN report on the State of the World's cities 2008/2009 focuses specifically on creating harmonious urban environments (UN-HABITAT 2008). Cities are becoming more unequal, particularly in the Global South, where wealth and poverty co-exist with slums often situated in close proximity to high-security (sometimes gated) privately owned residential communities. Latin America and the Caribbean, closely followed by southern Africa, are among the most unequal urban environments in which to live.

Although Europe, Canada and Australia have much lower disparities, inequalities in the Global North are still prevalent, and are especially demonstrated among US cities. Therefore we take an international perspective for exploring the causes and effects of social inequalities in cities.

Not only are the cities of the Global South growing at the fastest rates and hosting the greatest inequalities, they are also home to relatively youthful populations. This means that the experience of growing up in cities is by no means uniform for all young people as their lives are shaped by the social, economic and political processes that impose conditions on access to resources and livelihood opportunities, with the majority experiencing the negative side of inequality and poverty. The existence of stark inequalities creates divisions in the ways in which young people, both within and between countries, experience the physical fabric and social context of living in urban areas. In this chapter we explore the causes and effects of these inequalities, focusing in particular on the problems of growing up poor. We begin by focusing on the structural causes of social inequalities and what it means to live in poverty before providing examples of how young people's lives are impacted by processes that can enhance these inequalities.

Causes of social inequalities in cities

An enduring characteristic of cities in capitalist societies is their unequal division of space. The question of how these divisions are produced and to what extent they are a reflection of social dynamics or a product of the organisation of urban space itself has occupied urban analysts for many years (see Smith 1999). In addition, the drive for cities in the South to become players in the global market, with strong infrastructure, transport and economic dominance as well as political stability, is juxtaposed with a backdrop of poverty, acutely highlighting the spatial divisions of inequality. Cities such as Johannesburg and Mexico City have secured global city status alongside immense inequalities. As Scott *et al.* (2001) point out, this creates cities of two extremes where massive poor communities living in shanty-towns and slums exist on the one hand beside spacious resourced communities of the wealthy on the other.

At a localised level within cities much of the work exploring why divisions exist has focused on the connections between class, race, ethnicity and residential segregation (Smith 1988; Harvey 1973; Massey and Eggers 1990; White 1999; Iceland and Wilkes 2006), while recent work has also drawn attention to the gendering of urban inequalities (Bondi 1998; Peake 1999; Bondi and Rose

2003; Klodawsky 2006) and to processes of inclusion and exclusion related to sexuality (Bell *et al.* 1994; Bell and Valentine 1995; Valentine 1995). The role of age has only received scant attention by comparison. Knox and Pinch show how demographic attributes such as age and family structure are related to patterns in the urban landscape. They emphasise the 'tendency for certain household types to occupy particular niches within the urban fabric' (2006: 75). Yet the authors only see a loose connection between age, family status and differences in socio-economic status. In this chapter, we seek to challenge this view and to demonstrate how tightly interwoven socio-spatial inequalities are with age.

Children and young people are all too often eclipsed in household-based research, which tends to focus on adult income earners and often exclusively on an assumed male head of household:

> children are accounted for in terms of their parents' economic situation, and they are thus split up in accordance with criteria that do not characterize their own life conditions. The material conditions of children as a category are thus hidden, and much more so if the 'family' is the unit of observation.
>
> (Qvortrup 1994: 16)

Yet, children under the age of 18 are disproportionately affected by poverty and there is a close connection between socio-economic status and the number of children in a household. This means that there is a strong relationship between age, family composition and socio-economic status that warrants closer investigation.

When examining children and youth's social well-being in the city, we need to understand a wide range of intersecting factors: their specific age-related position, their household's composition, wealth and status, the effects of urban dynamics such as the housing market, social differences and inequalities based on gender, race and ethnicity, and the effects of different political and economic systems. The social well-being of children and youth is affected by restrictions of their citizenship rights and earning potentials, by the economic position of their household and by a large number of other factors that may increase social marginality, such as racism, disability, the stigmatisation of place, crime and fear of crime, low quality education, poor childcare facilities and other lacking infrastructures and amenities. In addition, growing wealth disparities across urban areas disproportionately create unrealistic expectations for young people living in poverty. Surrounded by influences from the media regarding middle-class global youth cultures, young people are encouraged to desire commodities and lifestyles that they cannot afford (Leitchy 1995). In some circumstances this draws poor youth into criminal activity as they seek to emulate the fashion styles, and acquire commodities such as mobile phones, they see in the cities.

Inequality and urban poverty

Class, wealth and gentrification

A journey through most cities in the Global North and South reveals highly differential living conditions reflected in the quality and appearance of houses, the local infrastructure, the presence or absence of managed green spaces, and the lack or abundance of amenities, shops, restaurants, etc. Poverty and wealth are often concentrated in certain areas of the city and yet they can also be found in very close proximity, such as in inner cities characterised by expensive hotels, shopping centres and office blocks on the one hand and some of the worst quality housing and homelessness on the other (Goodwin 1995).

Social geographers have identified a number of reasons why cities in capitalist societies display such stark wealth divisions. Although they highlight how the living conditions in 'poor places' form part of a cycle of poverty from which residents often struggle to escape, they argue that place characteristics in themselves are an insufficient explanation. Rather, we need to examine 'a variety of different economic, political, social and cultural processes operating on different but overlapping "surfaces", some of which are national or even international in their extent' (McCormick and Philo 1995: 7). For structuralist geographers, divisions of wealth and poverty are endemic to capitalist societies with their spatial concentrations of capital and resources (ibid.). Changes in the flows of capital and in the 'spatial division of labour' (Massey 1984, 1995a) affect the economic role of cities and of particular groups within them. This was evident across North America and Western Europe with the decline of manufacturing in the late 1970s and early 1980s. Cities based on manufacturing or mining lost much of their economic potential and those previously employed in the sector struggled to find new employment. Changes at the international level thus directly affected the livelihoods of individuals and families.

Other factors leading to uneven divisions of wealth within the city are property markets, private investment decisions, mortgage lending policies, regulations of the rental market and the decisions of urban 'managers' responsible for planning and the redistribution of resources through political means (Smith 1988; Harvey 1973; Harvey and Chaterjee 1974). Changes in the property market, further, affect not only house prices and rental levels, but also the quality and type of available housing and the social structure of urban quarters. Where elite groups, such as those working in the cultural industries, join forces with property speculators, 'gentrification' often results. Gentrification has been defined as the process of middle-class residents moving into predominantly working-class areas in search of alternative lifestyles, lower housing costs and the possibility of gains through appreciating property values. While helping to regenerate areas, these investments tend to be followed by increases in property and rental prices that make housing and services in these areas unaffordable for poorer residents (Smith 1997; Hamnett 1991; Ley 1996; Lees 1994). Figures 3.1 and 3.2 offer illustrations of this process. Gentrification has, in some cities, been accompanied by more aggressive 'revanchist' policies (Smith 1996) of evicting and moving on residents that are deemed 'undesirable' and contrary to the image of a newly 'revitalised' urban quarter. Neil Smith (1996), a key writer on processes of gentrification and urban revanchism, connects these processes to the working of global capital and explains:

> Gentrification, and the redevelopment process of which it is part, is a systematic occurrence of late-capitalist urban development. Much as capitalism strives towards the annihilation of space by time, it also strives more and more to produce a differentiated space as a means to its own survival … The so-called renaissance is advertised and sold as bringing benefits to everyone regardless of class, but available evidence suggests otherwise. For instance, according to the Annual Housing Survey conducted by the US Department of Housing and Urban Development, approximately 500,000 US households are displaced each year (Sumka 1979), which may amount to as many as 2 million people. Eighty-six percent of those households are displaced by private-market activity, and they are predominantly urban working class … Just as economic restructuring at other scales (in the form of plant closures, runaway shops, social service cuts, etc.) is carried out to the detriment of the working class, so too is the spatial aspect of restructuring at the urban scale: gentrification and redevelopment.
>
> (Smith 1996: 89)

For families with children, the effects of gentrification depend largely on income and class status. New services, such as improved playgrounds, nurseries, shops and child-friendly restaurants, may well be developed, but access to them is restricted to those who can afford them and often also regulated through codes of behaviour or 'cultural capital' (Lees 1994). Previously open and unregulated parks and urban squares may become fenced in and tightly surveilled, giving those who are able to use them a greater sense of security, while excluding other users who are seen as troublesome. There are often restrictions on the age of users and on the status of adults who can accompany children to playgrounds. Unsupervised groups of youth may find themselves locked out even if the area is promoted as child-friendly. The

Figure 3.1 Gentrification in Berlin, Germany. (Source: Kathrin Hörschelmann)

Figure 3.2 Gentrification in Cape Town, South Africa. (Source: Lorraine van Blerk)

result are homogenised spaces, where social diversity is reduced and 'safe' zones are created for middle-class children's play activities (see Figure 3.3) The learning task on page 49 encourages you to investigate these issues.

The redevelopment process can also have similar effects on a broader scale and is by no means something that is reserved for cities in the Global North. Box 3.1 uses the example of Cape Town, South Africa to demonstrate how the process of regeneration of the city centre impacted upon particular vulnerable groups.

Figure 3.3 Playground in a wealthy part of San Francisco, US. (Source: Kathrin Hörschelmann)

Box 3.1 **Regeneration in Cape Town's city centre**

South Africa is viewed as the economic leader on the African continent, with Cape Town quickly following Johannesburg as an international centre. For such cities to compete in the global market there is immense pressure for resources to be targeted into promoting business development, investment, retail and entertainment. In 1999 the Cape Town Partnership (CTP) was established to regenerate the city centre, previously awash with vagrants, street children and criminal activity and had resulted in businesses and residents relocating elsewhere. One of the CTP projects was the Central City Improvement District (CCID), established in 2000, to clean up the city centre by removing crime and dirt through large-scale employment of security services and street cleaners. The intention was to create a city centre that would be an attractive location for businesses. Since the inception of the CCID significant improvements have been made with massive private investment in the city and new businesses flocking to claim central city spaces.

Such investment has been at the expense of the poor. Although Cape Town's local government implemented pro-poor strategies, such as using local economic growth to create employment opportunities and alleviate poverty, these polices have often conflicted with the pro-growth agenda. This has heightened spatial and social polarisation in the city as investment continues to support the wealthy middle classes in the city centre, formerly white northern and southern suburb areas, while the poverty-stricken black African and coloured slum communities in south-eastern areas of Greater Cape Town are not benefiting from the investment which is diverting resources away from social provision in these communities. Although some effort was put into developing business investment in poor communities, businesses have avoided this in favour of wealthier locations where higher rates of return are expected.

The regeneration of the city centre has pushed social problems, once a visible reminder of the existing inequalities, out of the inner-city boundaries into more hidden poorer locations. Instead of addressing these problems as part of the regeneration, they have simply removed them from the public eye and placed them back within the already poverty-stricken communities. For example, homeless young people were, and still are in some instances, removed from the streets and taken home or to shelters in conjunction with disruptive practices to their livelihoods such as arrests for being unruly in public, being woken at night and having their blankets taken as well as being taken out of the city. This has cleaned up the inner-city streets but only displaced the social problems facing poor young people into locations where they are less likely to be noticed and supported.

Source: adapted from Lemanski (2007) and van Blerk (2011)

Learning task

Conduct a survey of two or more socio-economically different neighbourhoods. You may be able to use local statistical information to establish key indicators of well-being and house price trends in the areas. In your survey, consider:

- the ability of households with one or more children, and from different income groups, to afford adequate housing in the area
- the dominant type of accommodation
- the presence or absence of gardens, parks, traffic-calmed and pedestrianised zones, freely accessible playgrounds, childcare facilities, doctors' surgeries, public transport, shops, etc.
- the quality of parks, playgrounds, childcare facilities and other amenities
- signs of segregation by race and ethnicity
- signs of gentrification, or degradation, if applicable
- the 'feel' of the place
- representations of the neighbourhood: in the local media, by outsiders, by residents, especially children and youth.

What features can you identify that would promote or hamper the well-being of children and youth?

Race and ethnicity

As the learning task also alludes to, wealth divisions in the city are related to other forms of social exclusion, particularly racial discrimination. The political system of resource allocation, exclusionary policies by mortgage companies, racist reactions by residents, inclusion and exclusion from the labour market and the need for support structures within migrant communities are all factors that have been identified as playing a major role in the segregation of urban populations by race and ethnicity (Smith 1988; Platt 2002). White (1999) names four major types of social exclusion affecting ethnic minorities in European cities. These are exclusions through

- legal mechanisms
- the ideologies of 'othering'
- denying minorities access to social capital
- poverty and economic marginalisation.

Children of ethnic minority groups continue to be worst affected by poverty and social exclusion (Platt 2002). Thus, for the UK, Flaherty *et al.* (2004: 145) have stated that 36 per cent of Indian, 41 per cent of black Caribbean, 47 per cent of black non-Caribbean and 69 per cent of Pakistani and Bangladeshi children lived in households with an income below 60 per cent of the median (AHC), compared with 27 per cent of children in white households. In the US, the situation is similar with 63 per cent of American Indian, 61 per cent of Latino children and 60 per cent of black children living in low-income families, compared with 27 per cent of Asian and 26 per cent of white children (NCCP 2007). Thus, when analysing children's socio-economic position, it is not enough to conceptualise 'childhood' in terms of a structural entity of its own, as advocated by Qvortrup (1994) and Thorne (1987). While such an approach highlights mechanisms of exclusion based on age, it insufficiently explains why children from some ethnic and class backgrounds suffer significantly greater socio-economic exclusion than others. Ethnicity is a key factor in the social clustering of groups in the city. Yet, we need to remain cautious not to produce an image of the city as divided by impermeable boundaries of ethnicity, race and class. As Massey *et al.* (1999), Amin and Thrift (2002) and others have pointed out, recognising hybridity, interchange and flux is important if we seek to challenge racial inequalities.

Gender

In order to understand the causes and effects of poverty, we also need to consider how income inequalities are related to gender. Although there have been major shifts in favour of female employment in Northern cities over the past thirty years, women continue to earn less than men and are overrepresented among part-time, temporary employees (Massey 1994, McDowell and Sharp 1997). Women still carry out the majority of unpaid caring work and are more likely than men to look after children when families break down. The majority of single-parent households are headed by women and it is these households and their children that are at greatest risk of poverty. Gender also plays an important role in the experience of poverty among urban boys and girls. The latter will more frequently be expected to carry out household duties and caring roles, which circumscribe their leisure time and activity space even more than is already the case for boys growing up in poverty. In countries where girls are seen as a burden on family resources, they are also more likely to suffer neglect, abuse, engage in prostitution and be excluded from education and health care.

Box 3.2 **Neoliberalism**

Neoliberalism describes the market based way of organising the relation between state and economy that has dominated global trade and financial policies towards newly industrialised and less industrialised states since the debt crisis of the 1970s and has since been rolled out in the vast majority of industrialised countries, too. It is founded on neoclassical economic theory, particularly the belief that 'open, competitive, and unregulated markets, liberated from all forms of state interference, represent the optimal mechanism for economic development' (Brenner and Theodore, 2002: 2). State intervention in the economy is discouraged and the privatisation of formerly state-run enterprises promoted. Despite the critique of state intervention, however, 'coercive, disciplinary forms of state intervention in order to impose market rule upon all aspects of social life' (ibid.: 5) are part of neoliberal politics. Peck and Tickell (2002) distinguish between two interrelated phases: 'roll-back neoliberalism' and 'roll-out neoliberalism'. Under 'roll-back neoliberalism', previous welfare regimes and collective institutions are destructed and discredited (ibid.: 37), public assets are sold and public services are reduced, deregulated and contracted out to for-profit and non-profit agencies.

Under 'roll-out neoliberalism', 'neoliberalized state forms, modes of governance, and regulatory relations' are constructed and consolidated (Peck and Tickell 2002: 37). This involves the creation of new market regulations as well as socially interventionist policies and public-private initiatives, including welfare-to-work programmes and new surveillance and social control measures that seek to discipline, criminalise and control poor and marginalised social groups, while being couched in discourses of individualism, flexibility and entrepreneurialism. Participation in, and adaptation to, market needs and trends becomes the dominant way of advancing along social hierarchies, while cutbacks to welfare and public services explicitly and implicitly punish those who cannot or do not wish to participate in the new social and economic regime. Neoliberal policies have, instead of the assumed 'trickle-down' effect of wealth from the wealthy to the poor, led to increasing socio-economic inequalities and an erosion of social security for the most vulnerable and disadvantaged in society. As we have seen throughout this chapter, this affects children and youth disproportionately.

The introduction of neoliberal policies has had profound effects on cities. Not only have public services been eroded, but privatisation policies and cutbacks to social housing budgets have led to a reduction in affordable housing, increases in rental prices, and the sometimes forced removal of poorer residents from areas with profitable real estate. Cities are increasingly viewed as competing economic units in themselves with the primary objective of selling the city to potential investors and attracting 'desirable', skilled and affluent residents, which in turn is assumed to increase a city's competitiveness in a global market. Zero-tolerance policies, curfews, anti-drug laws, the policing of begging and the practice of 'moving on' homeless people, as well as the reduction in social housing available in desirable residential and inner-city districts have all been part of this effort to 'sanitise' and in return increase the attractiveness of cities to investors and elite workers, while those at the receiving end of these policies have received no or limited support to be able to tackle the underlying socio-economic problems that lead to income inequality, homelessness, substance abuse and high crime rates.

For a succinct, critical introduction to neoliberalism and a useful set of papers on neoliberal policies in different cities, see Aguirre *et al.* (2006).

A further factor we need to remember is the effect of changing political ideologies and strategies on wealth distribution. The neoliberal regimes inaugurated by the Reagan administration in the US and Margaret Thatcher's conservative government in the UK in the late 1970s/early 1980s, for instance, have led to a significant erosion of social rights with severe consequences for those plunged into poverty by the loss of employment at the same time. (Box 3.2 outlines neoliberalism in more detail.) Coupled with economic recession, this meant that in the UK, child poverty increased three-fold from 10 to 31 per cent between 1979 and 1999/2000, with a further peak at 35 per cent in 1998/1999 (Bradshaw 2002, 2003; Flaherty *et al.* 2004). A UNICEF (2000) study of 25 countries found that Russia, the US and Britain had the highest rates of child poverty. The lowest rates were noted for Northern European countries, where transfers like social and housing benefit significantly reduced child poverty (Bradshaw 2002). There has been a decrease in child poverty since 1997 in the UK, but almost one in three children remain in income poverty (Flaherty *et al.* 2004: 145), measured at below 60 per cent of median income.

The effects of living in poverty on young people's lives

A recent UNICEF (2007) report on child well-being in 21 industrialised countries showed that children in the Netherlands and in the four Scandinavian countries fared best overall, while the US and the UK occupied the bottom ranks. Although the Netherlands only achieved medium scores for children's material well-being, it ranked high in the areas of health and safety, education, relationships, behaviour and risk taking and, most of all, subjective well-being. This shows that indicators of material status are not enough to define child well-being. Norway, Sweden, Finland and Denmark ranked highest in the measurement of material well-being. Relative income poverty here is below five per cent and social transfers are high. Income poverty is not directly related to employment. Thus, the US is one of the countries with the lowest rates of unemployment for parents with children, yet it has the highest rate of child poverty of all. UNICEF concluded that

> [h]igher government spending on family and social benefits is associated with lower child poverty rates. No OECD country devoting 10% or more of GDP to social transfers has a child poverty rate higher than 10%. No country devoting less than 5% of GDP to social transfers has a child poverty rate of less than 15%.
>
> (UNICEF 2007: 5)

UNICEF was particularly concerned about the trend in industrialised countries of increasing social spending on pensions and health care at the expense of investment in children. Relations of power based on age and wealth are thus very clearly expressed and lead to a mortgaging of the future.

Thus, poverty is concentrated in families with children. In the UK, they

> make up 53 per cent of those in income poverty. Children are more likely than other groups to be poor, one-third more likely than the population as a whole in the UK. Some children, such as those in lone-parent households, from workless families and many from ethnic minority families, are more at risk of income poverty than others.
>
> (Flaherty *et al.* 2004: 145)

The situation is similar in the US, where 39 per cent of all children live in low-income families (NCCP 2007). In Eastern Europe and Russia, likewise, there has been a major increase in overall poverty levels since the political changes of the early 1990s and families with children have borne the brunt of these. According to a report by the European Children's Fund, there are around 50 million children living in poverty in Eastern Europe and the former Soviet Union today (Carter 2000):

> Analysis of the profile of Russian poverty prompts the conclusion that families with children make up the biggest group among the poor and are distinguished by a high risk and great depth of poverty. Extreme levels of risk and depth of poverty are characteristic of large families. Single-parent families are also a vulnerable group from the poverty point of view. At the same time, even the birth of a second child in a complete family increases the risk of poverty to 50% … Although children belong to the group with a high risk and considerable depth of poverty, the Russian system of monetary and in-kind forms of social support by the state is oriented on supporting the elderly.
>
> (Carter 2000: 56)

In cities such as Moscow and St Petersburg, where capital investments and economic growth are concentrated, young people without work (e.g. students and the unemployed) constitute one of the largest groups of the poor. These are cities already characterised by major income inequalities. The subsistence level here is higher than elsewhere in the country, with higher prices for housing, goods and services which also place households with working parents at great risk of poverty if they are employed in low-income jobs (Ovacharova and Popova 2005).

Single-parent families, particularly female-headed households, are also likely to experience higher rates of child poverty in other contexts. In Guatemala City, where more than half of female-headed households are poor, Hallman *et al.* (2005) note that rapid urbanisation (common throughout much of Latin America) has resulted in the decline of multigenerational households, which, when coupled with divorce and widowhood, leaves many women to run households on their own. In general, women have lower levels of education and more limited access to formal employment reducing their income earning potential. When coupled with the need to access childcare, this pushes many such families into situations of poverty. In many instances, this can perpetuate an intergenerational transmission of poverty, where poor people, relying primarily on the sale of their labour to meet daily needs, fail to invest their human capital in successive generations through skill or knowledge transfers (Kabeer and Mahmud 2009).

In spite of high levels of inequality present at regional levels, when considering the scale of child poverty internationally, this tends to mask the unprecedented levels of poverty faced by most children in sub-Saharan Africa. African cities have the greatest potential to grow, the most youthful populations and the highest rates of poverty, both in real terms and in terms of infrastructure and resources. For example, digital exclusion has exacerbated inequalities within sub-Saharan Africa and further resulted in the marginalisation of the region on a global scale (UN-HABITAT, 2008).

Figure 3.4 Poverty in Uganda. (Source: Lorraine van Blerk)

Different dimensions of poverty

When thinking about the effects of poverty on young people's lives, we immediately think about material deprivation. Images of children in squatter settlements with poor infrastructure come to mind. Poverty is partly about material deprivation, but also partly about experiences of social marginalisation. Gough and McGregor (2007) argue that material hardship is only one aspect of social exclusion and becomes a trap if other aspects of well-being are not taken into consideration (see also Searle 2008). For children, this includes education, health and living conditions, but also less tangible aspects such as forming positive, affirmative social relations and developing aspirations for the future.

As Goodwin (1995) has argued with reference to Townsend's (1979) research on urban poverty: 'The material and subjective aspects of deprivation are fused in an urban environment which highlights the multifaceted nature of poverty: part lack of income, part material deprivation, part social deprivation and part subjective experience' (Goodwin 1995: 75). Thus, we need to consider not only how material needs affect those living in poverty, but also how deprivation affects their 'engagement and participation in society' (Flaherty *et al.* 2004: 2). Flaherty *et al.* rightly critique the current focus on tackling the consequences rather than causes of poverty. They are sceptical of policy discourses which blame the poor for their own situation by emphasising the need to change attitudes and behaviour rather than tackling issues of material inequality: 'Approaches to poverty which do not acknowledge the centrality of enough disposable income to overcome market exclusion should be treated very sceptically, since they usually reflect the unwillingness of non-poor people to pay the price of redistributing resources' (2004: 27).

The case of child poverty makes this particularly obvious. Young people under the age of 18 only have limited citizenship rights and, in cities of the Global North, few opportunities to earn an independent living. They are structurally marginalised (Qvortrup 1994), having less access to resources given their status as children. In the Global South, they may be partial or sole breadwinners, although their access to resources and rights can be similarly constrained; for example children are often employed as they can be paid less and are not regulated by employment laws. In both situations, their current standard of living and their future prospects depend crucially on the income and social position of their household. Where social welfare is cut to a bare minimum, it affects particularly those least able to 'fend for themselves', who are also already excluded from political decision-making processes. For children and youth the consequences are severe and affect not only their current standard of living but also their future prospects. The long-term consequences for their social integration and inclusion are worrying indeed. They suffer not only from a lack of all but the most basic necessities, but also from poorer health, lower education, weaker social networks that could lead to future employment, stigmatisation, peer group exclusion, strained family relationships and greater fear of crime.

For the UK, Flaherty *et al.* (2004) have identified the following key impacts of growing up in poverty:

1. parents making significant sacrifices
2. stigma
3. poor child health and development
4. educational disadvantage
5. moderation of demands on parents' financial resources and lower familiarity with financial institutions
6. long-term effects such as higher rates of future unemployment and economic inactivity; higher risk of early childbearing for females; unrealistic ambitions.

To these we may add how poverty affects children and youth's emotional well-being. They may be at greater risk of bullying, of exclusion from peer group or school activities, worry about crime in their neighbourhood and experience tensions in their social and family relationships due to the strain that poverty places on them.

Learning task

Poverty is defined not only in absolute, but also in relative terms to reflect the fact that incomes measured in dollars are rarely a sufficient measure of the extent and experience of poverty in different places. Using statistical and, where available, qualitative data provided by organisations such as UNICEF, the WHO, and the UN (e.g. the UNDP Reports), as well as by academic researchers (such as those cited in this book), outline what you think the key defining characteristics are of poor places in different cities around the world and how *relative* poverty there affects children and youth. Are there common experiences and causes of poverty in the different places you have examined, and to what extent are they different?

Living in urban poverty: life in low-income neighbourhoods

Taking these dimensions of poverty into account, we turn now to explore life in low-income neighbourhoods in both the Global South and the Global North.

Low-income neighbourhoods in the Global South

Physical infrastructure is a key concern for children and youth growing up in low-income neighbourhoods in the Global South. The majority of poor children, however, are living in informal settlements, whether legally or illegally, with often limited or no access to formal standards of provision. Access to piped clean water, secure watertight shelter, electricity, sanitation and refuse removal are not given standards. Children may have greater access to the neighbourhoods outside their houses as daily life often spills out onto the streets for playing, cooking, eating, etc. This can enhance the social well-being of residents, including children, through greater interaction. Yet, Bartlett (1999) points out that children can be more seriously affected by the environmental consequences of living in poor neighbourhoods because of their greater susceptibility to disease and hazards, highlighting how many poor urban dwellers can only afford to live in overcrowded (often one or two rooms) accommodation. In addition, informal settlements are usually located on the outskirts of cities far from services and employment opportunities, and often on land that is least desirable for housing, for example where the risk of flooding, earthquakes or landslides is high. This can have serious social impacts on young children in terms of access to childcare provisions and in the time they spend away from their parents who may have to travel long distances to work. D'Souza (1997) notes that the physical conditions in squatter areas in Karachi, Pakistan resulted in a high incidence of communicable disease among under-fives related to overcrowding, poor ventilation and insanitary conditions. Similarly in the Philippines, Racelis and Aguirre (2002) highlight several areas where children (aged 8 to 12) felt negative impacts on their well-being. These included material resources such as access to potable water, a lack of refuse collection but also social resources such as high crime and violence and a lack of police presence and protection limiting social mobility in the area. As one young person notes:

> you hear of young girls who disappear at night, then find out the next day that they are already dead because they were raped. Their bodies are just thrown over the wall over there. So we are scared. We are not allowed to go out … But criminals are rarely caught … Plus the police are lazy. Often when they are called they don't respond.
>
> (Racelis and Aguirre 2002: 105ff)

Low-income neighbourhoods in the Global North

Living in poor neighbourhoods is not the reserve of those in the Global South, although many of the issues facing young people in slums and squatter settlements are at the extreme edge of urban poverty. Despite this, young people in cities of the Global North also experience inequality and the effects of poverty in many similar and distinct ways. The appearance and provision of urban facilities contributes to poor children and youth's experience of deprivation and marginalisation. Poverty is concentrated in areas with few private gardens, playgrounds or secure, managed open spaces, thus reducing children's access to safe outdoor play areas. Affordable leisure facilities may be few and far between or located too far from their home. The absence of a family car or the cost of public transport also affects children's access to such facilities. Youth clubs can play a crucial role for providing meeting spaces and a range of leisure activities, but they are often under-funded. High volumes of traffic may further affect children's safety and opportunities for outdoor play, while within the home they have less access to a room of their own. This is in addition to feeling less safe in their local area and to living in low-quality housing. Where crime levels are high, poor neighbourhoods are often encircled by both visible and less tangible signs of exclusion such as widespread CCTV monitoring, few transport links into the area (including taxi restriction zones), boarded-up houses and barbed wire fencing around derelict buildings and estates. These signs further increase the sense of isolation and exclusion that residents, including children, may suffer.

Research by Hörschelmann and Schäfer (2005, 2007) found that young people's activity spaces within the city were severely affected by social inequalities. Those from poorer neighbourhoods rarely ventured beyond a tightly circumscribed radius of home, school, youth club and best friend, while those from wealthier backgrounds and higher level schools used a wide urban area for their leisure activities and frequently visited the city centre with its shopping and leisure facilities. Although public transport was relatively inexpensive in the city of Leipzig, where the research was conducted, young people from poorer families moderated their behaviour because they *felt* excluded from the world of cosmopolitan consumption and because their restricted financial resources limited the extent to which they could actively participate in it.

Box 3.3 further discusses the effects of growing up in a low-income neighbourhood through the example of Braybrook, Australia. Here the effects of material deprivation can be felt alongside social exclusion.

Box 3.3 **Growing up in Braybrook, Australia**

Braybrook is a 488-hectare housing estate in the western suburbs of Melbourne, Australia. The majority of Braybrook housing consists of semi-detached and detached houses, and flats up to three storeys high, constructed from prefabricated concrete slabs and fibro cement sheeting. Lack of maintenance has been an issue in the estate since its development. Neighbourhoods such as Braybrook were originally build to house workers in factories that made up large quadrants of industry and commercial zones built during this time. By the 1970s, most of the industry had been relocated and the factories abandoned, further providing reasons for disinvestment in the residential areas.

By the mid-1990s the population of Braybrook was mixed although dominated by a two-thirds Anglo-European population, with high levels of poverty in the area. One-third of adult residents of working age were unemployed and of those with incomes, half earned less than 50 per cent of the national average wage. Of the Braybrook unemployed, one-quarter were single-parent families and 36 per cent lived in rented public housing. In addition to the decaying fabric of the area there is a stigma attached to living in the estate as one young person reveals: 'Everyone thinks we're trash 'cause we live 'ere' (15-year-old boy).

In the research undertaken by Growing up in Cities, the young people in Braybrook described the neighbourhood as boring and dangerous, although this could be relative to neighbourhood experiences, as Sam's story reveals:

> My street is boring and quiet and 'nothin' much 'appens except well yesterday when a firebomb was thrown from a passing car at the house next door. I saw it from my lounge room window. And do you know it only took the fire brigade 15 minutes to arrive but the police didn't come for over 30 minutes. (Sam, 15 year old)

Young people in the study were asked to construct maps of their normal activity space to illustrate the extent of their roaming range. Their use of the neighbourhood environment generally included no more than one or two blocks from their homes. During mapping exercises, they described places in the environment that they avoided, such as certain streets where house occupants were identified as potentially dangerous or reserves where syringes were likely to be found, 'Parks are dangerous ... because sometimes you find needles ... I don't go there'. (1997)

Streets were recognised by one-third of the young people as the place where they felt most in danger. Both boys and girls listed drugs, alcohol, and physical and verbal abuse as the primary cause of fear in the streets. This was due to adults or adult activities (drug taking, drunkenness, policing). For girls, verbal abuse was normally related to incidents of sexual harassment, and for young people from non-English-speaking backgrounds, it was racial abuse and discrimination.

There is not a lot of entertainment specifically for young people, as they are not a section of society with great economic resources. With the privatisation of sport and entertainment facilities, their access has become even more limited. In addition, adults often complain if older children use playgrounds designed for the younger children. Competition over leisure resources means tension between adults and young people, and between groups of young people also exists.

Because of their age and the lack of physical resources, most young people in Braybrook had a limited capacity to find alternative spaces when tensions arose. In fact, very few young people knew or used facilities in neighbouring towns ... Most of the young people's parents are either very poor, working long hours, or single, and therefore unable to provide their children with opportunities for extended discovery. Consequently, young people in Braybrook are growing up

> with limited environmental experiences ... As a consequence of the tension in their own neighbourhood and the lack of exploration into other neighbourhoods, children retreat to what is becoming teenage space – the lounge room or den.
>
> Source: Malone and Hasluck (2002: 83ff)

The outcomes of poverty on young people's lives

Although there are many different processes that exacerbate situations of poverty for children and youth, in this section we draw on some key examples to illustrate the diverse nature of urban poverty and how poverty can produce particular conditions such as children's work/labour, mobility and migration and homelessness. These outcomes often create situations of heightened inequality yet at the same time can alternatively offer escape from situations of poverty or enhance general well-being in the immediate term.

As we discuss, these diverse situations of poverty we will explore the outcomes for children and youth through the lens of transitions to adulthood. This is helpful in this context as poverty often exacerbates many of the areas often identified as part of young people's transitions to adulthood and often at a younger age. So, where traditionally leaving home, finding work, leaving education were considered part of transitions to adulthood, enabling young people to carve a pathway from dependence to independence (Bowlby *et al.* 1998; McDowell 2002), this pathway is now becoming increasingly fractured and diverse resulting in many paths to independence (Jones 2002). This diversity is in part exacerbated by the impacts of poverty. For example, transitions to adulthood are different where the impacts of economic restructuring and unemployment make it difficult for young people to find steady work (Jeffery 2010), where leaving home results in homelessness or where employment is found only in illegal, difficult or dangerous sectors (van Blerk 2008), where religion dictates the type of work you do or who you marry (Hopkins 2006) and transitions can be about interdependence rather than independence (Punch 2002), where young people may use their earnings to support their parents. In the remainder of this chapter, as we explore young people's work, migration and homelessness as outcomes of poverty, we will consider the ways in which these processes have both positive and negative impacts on the lives of children and youth.

Working in the city

Engaging in paid employment is something that the majority of young people will, at some point, participate in regardless of locality, situation, gender, race or status, although the types of work may vary according to these different identity positions. For some this will be helping the household with chores, for others it will be paid work to help supplement personal or family income, and for others still they will be the main breadwinner in the family. For those working directly for some form of income their work is more often firmly located in the informal sector, generally receiving little support or protection from unions or agencies, such as the child buskers in Figure 3.5. McDowell (2004) notes that, in the US and the UK, increasing numbers of young women, members of ethnic minority groups, students and children engage in paid work in a range of low-paid casual and temporary positions that are contractually varied. Thus, the reality for many children and youth living in situations of poverty is that work forms part of their daily routines and that often this work is poorly paid, morally demeaning and/or hazardous.

There is some debate regarding whether it is appropriate for children to work, particularly in societies where many young people engage in apprenticeship work from a young age or where cultural traditions reserve some tasks for children as part of their obligations in supporting the family (Kielland and Tovo 2006). This has led to a distinction in the literature between child work, which constitutes all

Figure 3.5 Children working as buskers in Gdansk, Poland. (Source: Kathrin Hörschelmann)

work-related tasks undertaken by children including household chores, and child labour, which refers to activities that are harmful to children and should be eliminated. The International Labour Office is actively working to reduce child labour through their International Programme on the Elimination of Child Labour (IPEC). IPEC is the ILO's 'operational arm' in the fight against child labour and works to strengthen national capacities in developing prevention activities and supporting children to exit particularly intolerable forms of work (Lansky 1997). In addition, the ILO (2006) has separated out some activities as especially dangerous under IPEC Convention 182 regarding the elimination of the worst forms of child labour. These include all forms of slavery, bonded labour, commercial sexual exploitation of children and the use of children in crime including trafficking of drugs.

Children working in cities

A greater percentage of working children are located in rural areas, where most take part in traditional family activities which often include agricultural work and household chores. However, as previously noted, many children who migrate into urban areas (independently and with their families) also end up working. A study undertaken in Nairobi, Kenya highlighted that the difficulties facing urban child workers can be masked by statistics that rarely show the vast inequalities across cities. Mugisha (2006) found that children living in Nairobi's illegal slums experience deep-rooted poverty that often results in their working under more intolerable labour conditions than their rural counterparts. The children in this study were engaging in prostitution and selling drugs or snacks outside bars and entertainment establishments. They were likely to work longer hours, often late into the night.

Figure 3.6 Market work, Malawi. (Source: Lorraine van Blerk)

Figure 3.7 Head loading, Malawi. (Source: Lorraine van Blerk)

Similarly in East Asia, young urban workers have been found to engage in activities that are dangerous to their health and well-being. For example, 80 per cent of young sex workers in Mumbai, India were found to have contracted sexually transmitted diseases (Bartlett *et al.* 1999). Most urban working children, however, undertake work as part of family businesses in small micro-enterprises selling homemade food items, collecting scrap or working in transport. Some example of this type of family work are represented in Figures 3.6 and 3.7. In some of these occupations, such as apprenticeship work including working as a mechanic, hazards could be avoided through the provision of protective clothing (Kielland and Tovo 2006). For example, in Bangladesh children were found to be involved in welding and working with acids for battery making. They were working without proper protective glasses and gloves (Bartlett *et al.* 1999). It is this lack of protection afforded to most urban child workers that makes their work dangerous. Dyer (2007) notes that in Yemen 87 per cent of working children work for families rather than wages. This means they are not protected by the Yemini Labour Act which does not consider helping out in family businesses as work (Dyer 2007).

The reasons for child labour in urban areas are numerous and although they generally stem from poverty, cultural factors play a major role. Box 3.4 illustrates this, using Bangladesh as a case study. Similarly, CINI-ASHA (2003) found that cultural values were important for Indian children engaging in work from Calcutta's slums. Many of the children in the study did not attend school and when the NGO CINI-ASHA became involved in finding them places in formal school their families were able to reduce expenditure on luxury items and festivals in order to compensate for the reduction in/or lack of children's earnings. Further, despite allowing their children to attend school most of the fathers felt that it was not appropriate and that children should work. This demonstrates that poverty is not always the main determinant in child labour but in fact only one determinant, with cultural values and traditions also influencing how and why children work. This has resulted in the debate on child labour focusing on the elimination of the worst forms of child labour.

Box 3.4 **Factors influencing children's work in Dhaka, Bangladesh**

The most common work activities for boys in urban Bangladesh are shop work, street selling and garment work, while most girls involved in paid work are employed in garment factories. Boys become more involved in work outside the home as they become older (over the age of 12) while the opposite is true for girls. Although an association was found between lower household income and children in paid work, and household poverty was the most frequent response from parents for why their children work, Delap (2001, 2000) found that there are other cultural factors that determine who works and in what occupations. Boys' paid work and girls' housework are commonly not essential to household survival, as adult members are usually available to replace their labour. Despite this, in Delap's (2001) study almost 60 per cent of boys engaging in paid work came from households where an adult did not work. This could be explained by adherence to purdah, or female seclusion. This also meant that teenage girls were often the last resort for a household and would only be employed outside the home when all others, including female adults had gone out to work. This also explains the type of work that young people undertake. Boys are more likely to work in occupations that require them to move around public areas, including selling, while girls work in enclosed spaces such as the home or factory. The same was true for household chores, with boys undertaking chores that took them out of the house including collecting firewood and shopping. The age at which children engaged in work also had cultural implications with fathers stating that it was not good for an older child to be idle, and this was often a reason for the high number of boys over the age of 12 in employment.

Shocks, such as the 1998 floods which badly affected those living in Dhaka's slums, were also noted to affect children's involvement in both paid and unpaid work. During the floods many adults in households were unable to work due to illness or because their places of work were flooded, resulting in households relying more heavily on the income earned by children. Children felt valued by their contribution to household survival and some took on extra work to help their families. However, this placed extra strain on young people who had to work long hours often placing themselves in danger.

Source: Delap (2001, 2000)

The worst forms of child labour

Cities are also host to numerous girls who are employed in domestic service. Throughout Africa this is happening as only the very rich have adult servants (Kielland and Tovo 2006). Here child labour is hidden as girls often work for no wages, sometimes for relatives, but they can be subject to working long hours and face severe isolation if they are not able to return home for visits frequently. Sometimes domestic workers are subject to sexual abuse from their employers and this can be one of the paths to prostitution. For girls, and some boys, sex work is one of the most hazardous types of work to engage in. Box 3.5 describes the situation of Ethiopia's bar girls and highlights the exploitative nature of their work.

It is important to remember that children are engaged in labour activities throughout all parts of the world, not just the Global South. Eastern Europe has long been associated with child prostitution with the region hailed as a 'sex paradise' where customers have easy access to young children. Escalating poverty and the infiltration of Western commodities has also seen young people drop out of school to engage in black-market selling activities (Zouev 1998). Furthermore, in some rundown urban neighbourhoods in UK cities, children as young as 12 are involved in selling drugs, working as runners and spotters for adult dealers. Growing up in deprived areas, young people are attracted by the dealer's disposable income and seek to emulate them. Teenagers are missing school to work and some are so desperate to get into the trade that they initially offer to work for free (Campbell 2005).

Box 3.5 **Sex work in Ethiopia**

There are numerous ways in which poverty affects young people's life choices and results in engagement in sex work as a means of livelihood. In Ethiopia young people engage in sex work in different ways – on the streets, in bars and red-light districts. Although street boys do engage in transactional sex, this is mostly opportunistic and girls make up the majority of young sex workers who actively solicit their services. There are three ways in which poverty resulted in participating in sex work for the girls in this research. First, many girls were attracted to the cities from rural areas having been promised well-paid jobs or because they are attracted by the bright lights of the city expecting to earn enough to support their families back home. For these girls, they quickly realise that opportunities for poorly educated rural girls are limited, only able to find work as maids or waitresses. These occupations carry with them difficulties of their own. Girls who found jobs as maids often recounted stories of abuse, both physical and sexual, from the owners of the homes they worked in which resulted in them leaving and entering sex work as a last resort. For those who took jobs as waitresses in coffee shops and bars the situation was not much better. There is no pay for this work and income can only be earned through tips and providing additional sexual services to customers. Others reported running away from early marriage (as young as 12 or 14 in some cases) or *t'ilf* marriages because their parents could no longer support them and would be able to use dowry payments as income. (*T'ilf* is marriage by abduction, which sometimes involves rape. Once a girl has been taken in this way her parents will not allow her to come home.) Often the girls reported being married to older men, whom they did not like. In addition, most of the girls (and some boys) living on the streets use transactional sex as part of their survival strategies. Most of the street girls had left home to avoid traditional practices such as genital cutting, because they had brought shame on the family by having sex outside marriage, or simply because of poverty in their home lives. In the latter case, some girls felt that as the eldest sibling in the household they should find alternative means to support themselves in order to relieve the burden on their parents. On the streets they live in desperate poverty which forces them to undertake begging and sex work for survival.

However, once involved in sex work there are very little options for young people to seek support. For many organisations, young people engaging in sex work is viewed as abuse rather than work, making it difficult for them to seek medical help, including contraception and other services. Most projects deal with older women (and definitely not boys) making this a particularly hidden occupation. This results in young people being disempowered in their transactions with customers and opening them up to other forms of exploitation and abuse. For example, the girls reported earning very little or no money with a very good evening's work making them only 30 birr (£1.80) from which they must pay half to the bar owner. They are also generally unable to negotiate condom use effectively, are tied to bars or red-light districts through debt to the owners, and on occasions are beaten and raped. Despite these negative consequences of engaging in sex work, many of the young people acknowledged that on some occasions it provided them with enough resources to send home remittances and offered them an alternative survival strategy to early marriage in their home districts.

Source: van Blerk (2007a, 2007b, 2011)

Youth and work in the city

Independence from parents (or in some cases generational interdependence) is one of the clear implications for youth entering into paid work, particularly where they are experiencing extreme poverty in their home situations. Yet current economic and social changes such as the impact of economic collapse at various levels across the globe and market restructuring have affected the ways in which youth are able to successfully negotiate 'reaching adulthood' through transitions to work (Jeffery 2010). The effect of global economic crises for youth's ability to secure employment can mean prolonged unemployment and uncertainty. Drawing on their work with working-class young men in urban centres in the UK and US, both McDowell (2003) and Katz (2004) illustrate the impacts of growing up in low-income working-class neighbourhoods for young school leavers. The lack of access to jobs in manufacturing through de-industrialisation and the disinvestment in surrounding neighbourhoods provide young people with a less-than-aspirational attitude. The jobs available to their fathers in the 1960s and 1970s are no longer there. Further the rising importance of school and post-school qualifications has meant that school leavers with limited passes are disadvantaged, resulting in a crisis of independence or as Jeffery (2010) calls it a permanent status of youth. This has negative consequences for young men's masculine identities as their inability to secure permanent work in their chosen occupation may have the knock-on effect of putting off reaching adult status through marriage or economic independence, whether this be in Northern India (Jeffery 2008), New York (Katz 2004) or Sheffield (McDowell 2003).

Youth's ability to pursue employment in the city can also be constrained by markers of social difference, such as gender and ethnicity, and the influence of parent–child relationships. Bowlby *et al.* (1998) note the relational aspect to entering work and the influence of parents in terms of choice of employment. They argue that gender, race, class and educational attainment equally impact on the types of work young people engage in. From their research with the Reading Pakistani population in the UK, they noted that young women were significantly constrained in terms of the type of work they were allowed to engage in and for whom, preferring their daughters not to have interaction with the non-Muslim community. Therefore the social networks necessary for young female Muslims to find employment were also constrained as the majority of their older female kin were not engaged in paid employment. In some circumstances, then, young people are constrained in their ability to access employment as an alternative to living in poverty. This can mean moving away from home to find alternative opportunities. Punch (2007) in Bolivia, Swanson (2010) in

Ecuador and Jeffery (2010) in India all discuss young people's mobility from rural areas into urban centres to access work, sometimes crossing national borders in the process. The city becomes a space of opportunity in such contexts, offering an alternative to rural impoverishment. Similarly migration for work has provided advantages for young people seeking access to alternative markets and lifestyles provided cross-culturally through au pairing. Au pairing is the process by which youth from across Europe are allowed to migrate for cultural advancement/work opportunities. In recent years this has resulted in many young people from impoverished Eastern European states seeking access to the cultural and educational capital contained in Western European cities. Au pairs are provided with a host family and obliged to undertake 25 hours of light housework and childcare per week in return for their accommodation, 'pocket money' and an opportunity to engage in language classes. Cox and Narula (2003) discuss au pairing in London and note that strict rules around family involvement, visitors, time off and blurred family/worker boundaries mean that many young people are constrained by their ability to negotiate the power relations with their host families affecting their ability to network and access better, more stable employment opportunities. In many parts of the Global South those that migrate into cities seeking work also remain in poorly paid informal jobs through lack of education and/or social networks.

We must also recognise the resourcefulness of youth in carving out their own transitions to adulthood and using all forms of work for creating independence/interdependence. While recognising that child labour might be considered exploitative, by focusing on youth there is evidence that even the worst forms of work can create independence (McDowell 2002). The ILO is actively encouraging youth employment in some sectors in an effort to reduce child labour through the Minimum Age Convention 138 (ILO 2007). Additionally, in some contexts exploitative work can be seen as having an empowering effect on young people. In South-East Asia, a number of studies have shown that young women migrate to the cities in search of employment in factories (Camacho 1999; Lauby and Stark 1998; Silvey, 2001). Silvey (2001) found that this was potentially problematic as many girls seeking work in Indonesia's factories found themselves unemployed due to the critical impact of wider economic processes on the collapse of factory employment. However, some of Silvey's participants chose to engage in sex work rather than return to their villages as a way of making successful lives in the city. In addition, Box 3.5 also demonstrates that Ethiopian sex workers acknowledge that their employment enables them to support their families and take on adult roles and responsibilities.

Mobility/migration and poverty

Migration to urban areas is an important aspect of many young people's lives, particularly in societies where they are influenced by the interconnectedness of (family) relations across space, thinking here particularly of the impact of labour migration on families in southern Africa (van Blerk and Ansell 2006). In the Global South, poverty is increasingly influencing young people's migration in complex ways and may be influenced by individual/family reactions to the impacts of much larger societal processes (Young 2004). This section will focus on different ways children and youth migrate to cities and explore what this means for their ability to respond to poverty in their lives. Migration for employment, as already alluded to briefly, family migration to slums and squatter areas, migration to relatives and migration to the streets all afford young people opportunities and constraints in terms of their transitions out of poverty.

Migration to cities

As we have pointed out in the previous section on work, youth are engaging in migration as part of their transitions to adulthood. In many circumstances this is migration to cities from impoverished rural areas in an effort to move out of poverty. Evidence from Sudan and the Ecuadorian Andes suggests that young people's work opportunities are changing in rural areas due to the impact of global restructuring on rural livelihoods as this may no longer provide knowledge and values important for their future (Katz 2004; Swanson 2005). For example, long-term structural impacts on households and families, such as the AIDS pandemic which has swept across much of sub-Saharan Africa, influences household and community poverty and has resulted in food insecurity for many in southern Africa. Gillespie and Kadiyala (2005) provide evidence to suggest that AIDS is damaging agricultural livelihoods, which is particularly likely to affect young people's futures and result in rural–urban migration for work. As Katz (2004) points out from work in rural Sudan, from the mid-1990s young men were beginning to migrate into urban areas in search of work. These changes were brought about by politico-economic changes in their villages and resulted in those who had no access to farm tenancy seeking labouring work outside the rural area. Similar reasons for migration for work have been found in Asia and Latin America where employment opportunities in rural areas, coupled with family poverty, have resulted in young people leaving their family homes. Research in the Philippines has, for instance, highlighted the role of young domestic workers in family survival. Here children, usually girls, move to Metro Manila in search of employment as a means for supporting their families (Camacho 1999; Lauby and Stark 1988). In the Latin American context, Punch (2002; 2007) found that eroding rural livelihoods due to a lack of available agricultural land for Bolivian young people who have not yet inherited from their parents, and few alternative paid employment opportunities, had resulted in urban migration. This formed part of young people's transitions to adulthood in an effort to be self-sustaining and assist their family. While in Mexico, Mutersbaugh (2002) found young people engaging in circular migration to urban areas to work, under often exploitative conditions in domestic service, as a method to attend (and pay for) city high schools. The strategy for migration in these contexts is to develop financial, educational and cultural capital that will help to foster a reduction in the impact of poverty on their lives (Figure 3.8).

Moving to relatives

Children's independent migration is not only for work purposes and in many cultural contexts children are often sent to urban areas to help relatives for a variety of reasons. This can form part of their transitions to adulthood and can include helping childless relatives with the housework or in order to access better educational opportunities. In such cases relatives are usually able to pay school fees and look after the children in return for their help around the house with chores. For example, Orellana *et al.*'s (2001) work discusses how African-American children engage in circular migration to help relatives and how children can act as social, emotional and economic 'linchpins' for households that span transnational borders. This type of reciprocity is more commonly acknowledged in southern Africa where the history of labour migration is such that families are often spatially dispersed, and it is not uncommon for children to be sent to live with urban relatives. Although this is not a new phenomenon, situations of poverty in many cases are exacerbated by the AIDS pandemic, and children's displacement is becoming more common for example in Brazil (Fontes *et al.* 1998), and southern Africa (Ansell and van Blerk 2004b). In such cases children's migration is used as a coping strategy for households affected by AIDS, particularly where income has been lost through the sickness or death of a breadwinner. Children may be sent to urban relatives to be cared for, to be able to attend school, to help care for a sick relative, or to help with household chores. In some instances this has the effect of bringing children out of material poverty, and can even foster support for transition through education, particularly where they move to live with better-off relatives, but can also exacerbate emotional and social poverty through the isolation of living in an unfamiliar environment.

Family migration

As we have so far noted, young people often migrate independently, however, they also move with families, sometimes as refugees but also when whole families feel under the strain of rural poverty this acts as a motivator for family migration. The stark inequalities that exist across cities (see the section on causes of inequality earlier in this chapter) often motivates rural to urban migration. In locations as diverse as Ethiopia and the Ecuadorian Andes, cultural, traditional and more recently globalised circumstances promote young people begging on the streets of large cities. In Ethiopia poverty in the northern regions coupled with a cultural-religious tradition of begging at particular times has resulted in families migrating to the cities. Similarly, Swanson (2005) notes that recently women and children have been joining the ranks of migrant men and heading to the city to beg or sell gum on the city streets in Ecuador. They are leaving poverty-stricken Calhuasi, where economic globalisation is disintegrating rural livelihoods and making their comparative poverty more acutely noticed. Although they have not necessarily become poorer (their poverty is based largely on a colonial history of racism and social exclusion), they have swapped agricultural work for informal street work as a method of improving their livelihoods.

In many cases migrant families that do not end up on the streets enter the informal housing market, often renting small rooms or constructing shacks in low income neighbourhoods and squatter settlements. Although the physical provision may be considered better than their rural dwellings in relation to access to services and resources, such areas are also associated with poor quality accommodation, overcrowding, poor maintenance and lack of access to infrastructure (Andreasen 1996; Watson and McCarthy 1998). In addition, migrants who are unable to connect quickly to other migrants from their own rural areas may experience social and economic discrimination including lack of security, teasing and insults related to their tenure and outsider status, as well as problems of conflict with landlords/ladies, who often live in the same compound as the tenants (Watson and McCarthy 1998; Ansell and van Blerk 2005; see Chapter 4 for a fuller discussion of the rental sector).

Street migration

Finally, one of the most high-profile aspects of children's independent migration into cities is street migration. Until recently, most studies equated street migration with poverty (Alexandrescu 1996; Scheper-Hughes and Hoffman 1998). Bar-On's (1997) description of the debilitating family environments that

Figure 3.8 La Nina, Mexico/ US border. (Source: Susan Mains)

result in children going to the streets as a micro-level crisis of poverty suggests that children leave when households, faced with extreme poverty, are unable to provide for their basic needs. However, Conticini and Hulme (2007) argue economic poverty as an explanation for street migration alone is insufficient and more recent work has sought to illustrate that street migration is a far more complex process (see also Young 2003).

Conticini and Hulme (2007), for example, argue that economic poverty is not causing only children to move to the street because many children living in situations of poverty do not engage in cityward migration. Instead, from their work in Bangladesh, they advocate that children go to the streets as a direct result of violence and abuse experienced in their homes and communities. Crucially then, it appears to be a complex combination of factors that result in a crisis of family poverty that are both economic and social. Here stress, violence, alcohol consumption and drug addiction are manifestations of economic poverty that seek to diminish family cohesion and resources (Dallape 1996; Matchinda 1999; Young 2004).

These micro-level causes of poverty are also related to wider structural conditions facing countries and continents with the effects of AIDS, civil insurgency and global economic restructuring. Within these debates it is important not to ignore children's agency in decisions to leave home and in particular to understand the complexity of their journeys to the street (Young 2004). Figure 3.9 illustrates that children's migration to the streets can be a fluid process and can be made up of several stages as children move first to shelters, relatives' homes or other small towns before finally heading to the city streets (Young 2004; Conticini and Hulme 2007). Additionally, such journeys are not necessarily final as young people may return home or stay in shelters for a while at various stages throughout their street career.

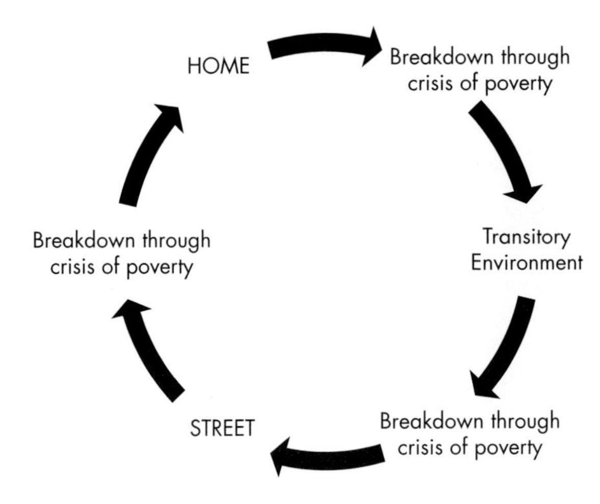

Figure 3.9 Journeys to the street. (Adapted from: Conticini and Hulme 2007: 220)

Homeless children: the public image of poverty on the streets

Once on the street, children and youth present a very public image of poverty and deprivation. From the 1980s and into the 1990s, when children's issues and rights came on to the international agenda with the 1979 UN International Year of the Child, street children have received much attention in policy and academic circles. Inflated numbers of street children in many cities, particularly in Latin America, helped to foster public outrage. For example, Tacon (1982) estimated the number of street children in Latin America to be 40 million. This led to discussion over defining 'who a street child is' rather than focusing on their needs as individuals. In part, this was an effort to remove children living on the streets, seen as a symbol of urban decay, from the public eye and create images of prosperous cities (and is still happening, as is outlined in Box 3.2). More recently, research acknowledges young people's agency in their decisions to live on the streets, suggesting that they are not destitute and in fact employ many skills and capacities in their complex survival strategies (Beazley 2000; Conticini 2005; Young 2003; van Blerk 2006). The city is a useful resource for street children utilising its many hidden places, both day and night, for a variety of activities that include sleeping, eating, playing and working. Figures 3.10 and 3.11 illustrate this showing street children gambling and sleeping on the streets in Uganda. Beazley (2002) notes that mobility around the city and excellent knowledge of its different locales provides lots of opportunities for earning money including begging on the streets, busking on the trains, shoe-shining, selling small items and helping in markets. Young (2003) illustrates how street children in Kampala, Uganda are marginalised in the city, pushed out of places for begging and sleeping by authorities, yet they resist this by taking over the streets, in some cases at night, through commanding a presence of stealing and drug-taking that others are fearful of. Yet the public image of street children is not always negative and many use their status as children to pick up informal work in markets and undertake other legitimate activities.

Changes in policy and practice for working with street children and youth have changed alongside perceptions of children in the streets from being out of place to having a place. Instead of viewing street children as deviant and corrupt or victims of society's evils and removing young people from their streets into prisons and shelters (Human Rights Watch 1997), children are now afforded agency and seen as having their own unique capacities, knowing what is best for their lives. Organisations now work to support street children with the choices they have made to seek their own survival on the streets. This approach sees handouts as encouraging more children to join the ranks of street children and instead projects now tend to offer support services and education as alternatives to providing direct support on the streets (such as food, clothing, etc., see also

Figure 3.10 Gambling, Uganda. (Source: Lorraine van Blerk)

Figure 3.11 Street children in Uganda. (Source: Lorraine van Blerk)

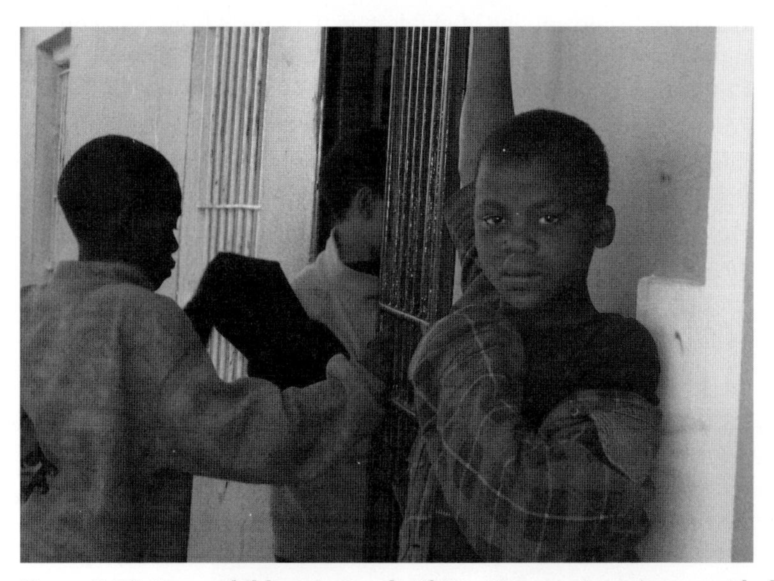

Figure 3.12 Street children in South Africa. (Source: Lorraine van Blerk)

Chapter 6, Box 6.5). Figure 3.12 shows street children accessing NGO facilities through their own initiative. In addition, NGOs moved away from seeing children on the streets as abandoned and institutionalising them, to an approach that sought to work through difficulties at home and set out to reunify street children with their families and by offering some support. However, although success has been achieved reuniting many children with their families, for others, particularly those who have stayed on the streets for a long time, the reunification often subsequently breaks down. The street has provided opportunities for friendship and work that are not now available in home communities (Conticini 2005; van Blerk 2011). When we return to think of transitions to adulthood or independence then, street children can be seen as one example of how young people have created alternative outcomes to family crises of poverty and (generally) moved away from their families, living (although mostly in a homeless context) and working (usually informally) independently on the streets.

Youth homelessness in the Global North

Young people living on the streets are not confined to the cities of the Global South and indeed youth homelessness is part of the fabric of many Western cities. It is an issue that is often neglected or hidden in the literature, with the focus of attention more on the youngest children who are generally moving to the city streets of the Global South. However, young people do increasingly migrate independently to Western cities. As Smith (1999) points out, throughout the 1990s Britain, France and Germany have all indicated rising numbers of homeless Western citizens on their city streets. Where previously the larger proportion of homeless people were refugees and migrants from Eastern Europe and Africa, this is no longer the case. It is unusual to find young children under the age of 16 homeless on the streets of cities in the West, with the majority of homeless youth (rough sleepers) in the UK being aged 16 or 17 years. However, Smith (1999) highlights that the reasons these young people give for leaving home are rather similar to homeless youth on the streets in other parts of the world with household conflict being the main reason, or to find work. Ruddick's (1996) text details the experiences of white homeless youth in Los Angeles, demonstrating how urban regulation forced young people to share space with other excluded groups and resulted in their engagement with illegal activities such as drug taking and crime. It is important to note, however, that social security systems in Europe have far greater capacity to support children and youth on the streets and therefore the magnitude of the problem cannot be compared.

In Eastern Europe, an increase in homelessness has accompanied the process of social, economic and political transformation over the last two decades. In Russia the numbers of homeless young people have grown substantially since the early 1990s and have caused a new moral panic in Russian society, where '[s]treet children are seen as evidence of the erosion of the very foundations of society' (Stevenson 2001: 532). For example, in Moscow, the number of children brought from the street to the special *militia* (police) reception centre for juvenile delinquents doubled between 1988 and 1998. This rapid growth in numbers is related to social processes resulting from the collapse of the Soviet social structure, including the breakdown of the Soviet welfare system and rising unemployment. In the absence of efficient social services and wide-reaching NGO provision, the *militia* continues to play a major role in dealing with street children (Stevenson 2001).

Explaining the persistence of young people's poverty

In order to explain why young people experience persistent poverty in urban areas and tend to be overrepresented amongst the poor, Danish sociologist Jens Qvortrup (1994, 1999) has argued for a systemic analysis, which focuses on both demographic and ideological factors. He proposes a shift in the distribution of responsibilities for children, moving away from family-centred ideologies that regard children as primarily a private matter:

> The ideological factor is that children were and remain a private matter; it is parents' responsibility to provide for children; the state may or may not be supporting in terms of family or children's policy, while other adults, organisations etc. have no responsibility. This is the gist of the family ideology, which because of its legal underpinnings actually functions as a kind of macro-structure in a very concrete sense. The demographic factor is a consequence of both decreasing fertility and prolonged longevity, which produce much more households without children under their roof. In Denmark, only 23 per cent of the total number are households with children, and given the family ideology the adults in these households are the only adults who must share their incomes with more than one or two persons, mainly children. More than three out of four households are thus allowed to use what they are earning for themselves. Since finally all children belong to a household in which the incomes must be shared between adults and children, but only one fourth of adults is in this position, it logically follows that the income of children relative to that of adults is bound to decrease.
>
> (Qvortrup 1999: 16)

Such an approach would share the responsibility for social reproduction more evenly, but would require a long-term commitment across adult society for the well-being of the future adult generation. It would mean a sea-change from neoliberal ideals of individual liberties and (alleged) self-reliance and a departure from prioritising the rights and needs of adults over those of children and youth.

A related but geographically wider perspective to explain the persistence of young people's poverty comes from the work of geographer Cindi Katz (2004).

Katz has analysed the effects of global economic restructuring on children's livelihoods, their social practices and their possible futures in rural Sudan and New York. We have pointed to her work throughout this chapter, but it warrants more detailed attention here. Her 'topography of global capitalism' shows the structural connections between processes that erode children's livelihoods and the bases of social reproduction in both of these places, exposing 'the derailing of children's futures in disparate places as common and simultaneous consequences of, among other things, particular and linked global economic processes' (Katz 2004: 156). Rather than focusing on seemingly 'local' processes alone, Katz has sought to demonstrate that the 'localized hollowing-out of children's prospects is traceable, at least in part, to global economic restructuring and the globalization of capitalist production … [G]lobal economic restructuring transforms the scale of uneven development, producing common effects in disparate local settings' (ibid.: 159). Thus, changes in local investment strategies affecting children's welfare, health and education are connected to global economic changes that are related to the dynamics of capital accumulation and the uneven spatial division of labour (Massey 1984, 1995a).

Conclusion: the effects of policy on reducing child poverty

The plural causes of child poverty mean that there can be no single successful strategy towards its reduction. State interventions within capitalist, market-oriented societies are unlikely to change the structural causes of poverty, but they can work towards the alleviation of its worst effects. Transfers, such as social benefits and tax credits, can reduce child poverty significantly (by as much as 80 per cent in Sweden and Finland: Bradshaw 2002), while state support for poor children and their families may also include subsidised childcare, greater investment in publicly funded education, improving access to health care, providing free or cheap local play and leisure activities, organising loan schemes and running family support centres (Cohen and Long 1998). Policies towards poverty reduction are increasingly influenced by an understanding of the wider effects and experiences of deprivation and by the need to listen to, and to improve participation of those most excluded from conventional political structures. This has, for instance, been the guiding principle for the recent UN initiative for child friendly cities, which has been implemented in cities across the world, and which we return to in Chapter 6.

The concept of social exclusion is today frequently used to express the wider range of factors impacting on children's well-being, of which material deprivation is only one. Social exclusion is a useful term for describing the combined effects of income poverty, political exclusion, social marginalisation, cultural discrimination, educational disadvantage and lower health. Yet, it is important not to forget that many aspects of social exclusion are a direct consequence of income inequality and that this issue needs to be tackled in tandem with other causes of social exclusion. The alternative is all too often practised by politicians who focus on the behaviour of the poor as a seeming solution to 'their' problems, while a major societal commitment to reducing inequality is sorely missing.

Despite this, if we take the example of child labour/work, there are some encouraging statistics that suggest improvement in poverty reduction for young people living in poverty in cities of the South. The ILO (2006) demonstrates that global awareness and implementation of policies to reduce child labour are working. They state that of the world's 317 million economically active children aged 5 to 17, 218 million are categorised as child labourers which is a significant reduction from the 250 million identified some ten years previously (Lloyd-Evans 2007). The implementation of IPEC's Convention 182 in 1999 to eliminate the worst forms of child labour has been seen to have the greatest impact on this reduction with fewer children now believed to be engaged in hazardous employment (Lloyd-Evans 2007). This, coupled with the almost exclusive ratification of the UN Convention on the Rights of the Child, has made it possible to have a worldwide concerted effort to create better living conditions for many children.

In conclusion, we have demonstrated in this chapter that young people living in cities do so in completely heterogenous ways, their experiences characterised by diversity and inequality. In all cities across continents, there are sharp divides between rich and poor that are spatial, social, economic and political. Poverty is experienced in different ways and in different contexts but the examples from the Global South demonstrate the extreme poverty facing many children and the processes at work that influence this.

Questions for discussion

1. What are the effects of the intersection of different forms of social inequality for children's well-being in different geographical contexts?
2. To what extent does the work that children do in cities challenge perceptions of both childhood and work?
3. How does migration help us understand young people's capabilities and capacities?

Suggested reading

Ansell, N. (2005) *Children, Youth and Development*, London: Routledge.

Bartlett, S., Hart, R., Satterthawaite, D., De La Barra, X. and Missair, A. (1999) *Cities For Children: Children's Rights, Poverty and Urban Management*, London: Earthscan.

Katz, C. (2004) *Growing Up Global: Economic Restructuring and Children's Everyday Lives*, Minneapolis: University of Minnesota Press, Chapter 6.

Useful websites

UN-HABITAT: http://www.unhabitat.org/

UNICEF State of the World's Children: http://www.unicef.org/sowc2011/

Consortium for street children: http://www.streetchildren.org.uk/

4 Growing up in the city

In this chapter we will:

- **explore how children and youth experience urban space in the city**
- **discuss the historical and cultural differences that influence the ways children and youth encounter urban space**
- **examine young people's varied experiences of living and playing in the city.**

Introduction

In this chapter we outline how children and youth experience the city in diverse ways. How they use urban space and how they are viewed within it has changed both over time (through the development of cities from pre- to post-modern) and over space (both within cities and between countries). Markers of social difference including gender, age, class, sexuality and ethnicity also influence the ways in which young people experience urban space through living, playing and working in the city. In this chapter, we begin by exploring how children and youth have made use of the city over time taking a historical perspective to changing urban society. Then, having examined young people's work in some depth in Chapter 3, we focus on two other key aspects of growing up in the city: living and playing. Throughout this chapter the ways in which markers of social difference intersect our discussions further demonstrate how growing up in the city is not only related to the significance of place but also the significance of social and cultural attributes.

Growing up in the city: historical change and experience in the city

In this section we offer an introduction and background to much of the contemporary experience of young people growing up in the city. By examining some of the radical developments that took place in Western societies during intense periods of modernisation and urbanisation, we can later draw links to the current ways in which young people live, work and play in our cities across the globe.

In our discussion of childhood in Chapter 2, we highlight the ways in which childhood has emerged as a concept over time. The historian Philippe Aries (1962) was perhaps the first to offer the suggestion that prior to the Middle Ages children were merely miniature reflections of adults engaging in daily tasks alongside their parents. Postman (1994) concludes that it was the onset of modernity and the technological revolutions, coupled with urban growth and movement towards urban living, that offered adults new ways of thinking about their children. The emergence of the capitalist regime in Europe and North America enabled children to be separated from participating in daily subsistence tasks for the production of their extended family households, no longer considered as economically viable (Valentine 1996a). The rise of the nuclear family and paid employment in urban centres meant children could be socialised within home and school environments as 'priceless' and in need of care and protection (see also Zelizer 1994). Those that deviated from this vision, predominately among the poor, required the control and correction of institutional organisations. Hence the emergence of the dualism of Apollonian and Dionysian childhoods that became so prevalent in the Victorian era (Jenks 2005; Wyness 2000; see Chapter 2 for more details on this).

The institutionalisation of urban childhoods that occurred at this time has been discussed in the works of Elizabeth Gagen, Laura Cameron and others who have explored the meanings and processes at work in the Playground Association of America and similar processes of schooling, training and correction in the UK. Drawing on work emerging in developmental psychology at the time, which in turn was based on Darwin's theory of evolution, childhood became understood as a series of stages that must be progressed through in order for children to emerge as fully competent adults ready to engage in society. In particular, rapid urbanisation and the influx of poor (immigrant) populations into slum neighbourhoods and city enclaves resulted in an image of wild unruly children roaming the streets or working in mines, mills and potteries as a cheap expendable labour force (see Davin 1996 for an account of growing up in London). Similarly, the extract from Joan Foster's (1997) account of children living in

Tyneside (see Box 4.1) was echoed across the Atlantic, for example in accounts by philanthropist Jacob Riis in the cities of North America.

Box 4.1 **Growing up in nineteenth-century Tyneside**

The nineteenth-century industrial revolution brought many advantages to the rapidly urbanising west. In Tyneside the boom created unprecedented employment opportunities and growth. In some areas, such as Elswick, terraced housing was built to accommodate the rapidly rising workforce. However, for those who were unskilled, 'the labouring poor', the only homes available were in the already overcrowded slums, such as in the Quayside and Sandgate areas. Here lived many of Newcastle's children. The poverty and deprivation were amongst the worst in the country. Children of the poor were an extra labour force that was cheap, expendable and unregulated offering advantages to factory employers that provided work in the textile mills, potteries and mines of the North East of England, no doubt replicated in many other cities across emerging capitalist Europe. In addition to the appalling working conditions, poor children inevitably lived in cramped, disease-ridden slum housing. The following quotation comes from an 1845 Report on the Sanitary Conditions of the North East and vividly describes the conditions in Newcastle' s Quayside and Sandgate areas, aptly termed the 'fever' districts:

> The streets most densely populated by the humbler classes are a mass of filth where the direct rays of the sun never reach … To take a single example of one of the more extreme cases shown to me when visiting them during the day, a room was noticed with scarcely any furniture and in which there were two children of two and three years of age absolutely naked, except for a little straw to protect them from the cold, and in which they could not have been discovered in the darkness if they had not been heard to cry.

Similar conditions in the poor urban neighbourhoods across England and Wales led to high mortality rates among children, which in 1851 meant that only 522 children for every 1,000 born reached the age of five. In his annual reports of the time Dr Thomas Barnardo commented on the appalling conditions faced by many children of the poor, highlighting epidemic diseases such as whooping cough, scarlet fever, influenza, diphtheria, measles, typhoid and tuberculosis as child killers, endemic in the impoverished neighbourhoods in which they lived.

Source: adapted from Foster (1997)

The response to these conditions could be seen in the increasing number and frequency of mental health issues. Gagen (2006) highlights that the newly emergent psychological theory connected mental ill-health to the effects of urban development on human character and the reforms placed on children were an attempt to order social life so that urban space did not disrupt the psyche of their future adult selves. Progressive education and juvenile crime reformers drew on the psychological theories of Stanley Hall who sought to connect mental and physical being. He argued that good character and morality was directly linked to muscular development, and therefore to instil moral character in America's future adults a programme of sport and exercise should be embarked on for today's young people. Reformers introduced regular short periods of physical activity into the school curriculum and set up a series of institutions from juvenile reformatories to playgrounds where activities were developed to facilitate the correct development of masculine and feminine ideals in a process to save the nation's children (Gagen 2000a). Playground reform was specifically introduced for immigrant (also read poor) children to remove them from the streets and place them in corrective environments. Gender was also a key factor in playground reform. Organisations socialised girls and boys in different ways, preparing boys for their role as adult men, as workers for the nation, and girls for a future of child bearing and homemaking. For this reason, girls over the age of 12 were not given separate playgrounds and structured programmes as part of the reforms, but instead remained in the playgrounds of the younger children as helpers. Boys over the age of 12 routinely engaged in sporting activities to develop their physical, and therefore moral, competencies. The public nature of the playgrounds enabled the transformation of young people to be seen by the public, and at the same time contained and separated young people from the street (Gagen 2000a, 2000b, 2006).

In the UK, the Malting House project in Cambridge in the early 1900s was a similar example of this (Cameron 2006). Here, much younger children participated in nursery education that used psychoanalytical theories to ensure the correct development of model citizens.

Although much less urbanised, the colonial peripheries were not completely absent from discussions of young people growing up in urban places. Lagos, in Nigeria, one of the earlier examples of African urbanisation, showed signs of child destitution. Fourchard (2006) notes that juvenile delinquency was similarly highlighted as a product of poverty and rapid urbanisation in 1920s Lagos, where immigrants swelled the impoverished districts of the city, and with the colonial administration of the time echoing many of the concerns noticed in Europe and North America. However, the spotlight on the problematic conditions of growing up poor in the Global South has more recently received attention as urbanisation progressed in the 1960s–1980s across much of the Global South.

Turning to examine more contemporary experiences of growing up in cities, many similar processes can be seen to impact on the lives of young people: poverty, institutionalisation, housing, play and work still remain central to the experiences of growing up in the city. The images of children as future adults that impacted so on the policies of nineteenth-century reform can be seen to prevail in popular imagination and the policies used to limit young people's interaction with urban space today. Despite this, there are also glimpses of young people engaging their competencies and individuality in the ways in which they experience their urban environments. In the remainder of the chapter, we turn to explore two key aspects of growing up in the city: living and playing. Of course, there are many other diverse ways in which young people experience growing up in the city, which are not part of this chapter but are discussed in other places throughout this book (see Chapters 3 and 5).

Living in the city

In this section, we discuss the different types of experiences young people have growing up in the city based on the neighbourhoods they live in, the type of housing tenure their families can afford, and how their experiences differ related to gender, poverty and other markers of social difference. Housing tenure and neighbourhood characteristics, including the social relations therein, feature heavily in accounts about what makes a good urban environment for bringing up children. The impact of the fabric of their surroundings is however more often missing in research accounts of children's lives, yet this has major implications for how they variously experience the city.

The generally high density of urban living results in a vast array of different types of communities co-existing in close proximity. The experiences of growing up in a particular city can depend very much on a variety of family and personal characteristics. The experiences of the wealthy in one city may be spatially proximal to those of the poor but worlds apart in terms of social, economic and cultural experience. Therefore, given the diversity of city housing stock, the area where it is located and the type of tenure all make for a vast array of experiences of living in urban environments. It is difficult to encapsulate all these experiences in this section of the chapter but the theme of diversity is central to exploring what it means to live in the city as a young person. By discussing the experiences of children and youth in a number of different settings across the world, we also demonstrate the significance of place in terms of growing up urban.

There are many analysing principles when considering young people's experience of living in the city that include gender, class, age, sexuality, ethnicity and it is our intention here that this diversity will filter through the following discussion, which separates out wealth as an organising principle for highlighting how young people's experiences differ.

Growing up rich in cities

Nowhere do the rich and poor rub shoulders more explicitly than in cities, despite their experiences being vastly different. Rich well-serviced neighbourhoods and gated communities are often situated close to dense slum areas or run-down social housing estates and squatter communities (UN-HABITAT 2008). More often the experiences of children growing up in poor neighbourhoods is documented and discussed (see Chapter 3 for a fuller discussion of this), although recently more attention has highlighted the ways in which the less poor experience the city.

The ways in which cities of the Global North can be seen to be places where children and youth 'live' could be debated. For instance, the construction of the rural idyll as the place to bring up children and the associated fears of urban environments for children, such as road safety and crime, have conjured up an image of the city as not a place for children and indeed scholars have coined the phrase 'out of place' when discussing the location of young people in city environments. 'Out of placeness' (Connolly and Ennew 1996; Sibley 1995) has been utilised to refer to children publicly visible in the city, which is borne out of theorisations around children's competencies and the corruption of the city by disreputable groups. Essentially, unaccompanied children visible on the urban landscape have been more associated with Dionysian images of childhood, resulting in the middle classes removing their children from such environments.

Gated communities

One response has been the development of secure enclaves where children are considered protected and safe, and often referred to as 'gated communities' or 'gated developments' (see Box 4.2).

Box 4.2 **Gated communities**

A gated community is a residential development where access is controlled, usually by security guards and private surveillance, and where fences or walls enclose houses, streets and associated amenities. Such privately built housing estates are often located on the edge of large cities and are the reserve of the middle and upper middle classes. More recently, gated communities have become prevalent in parts of Asia, South America, and South Africa as well as the more established communities in North America. The rise of such communities is based on a discourse of urban fear and by positioning security within, and boundaries around, specific enclaves has the aim of excluding undesirable elements of the city. In addition, Low (2001) argues this has the effect of increasing residential segregation (along race/class/ethnicity lines) more permanently in the urban landscape.

Source: adapted from Low (2001) and Alvarez-Rivadulla (2007)

The creation of wealthy, generally white, middle-class suburbs and particularly gated communities has developed at an expanding rate over the last few decades since the rise of retirement communities in middle class America. As Alvarez-Rivadulla (2007) states, they emerged as part of the housing boom in 1970s America but can now been seen in cities as diverse as Johannesburg, Beijing and Buenos Aires. Pow and Kong (2007) reveal the emerging divisions along class lines among the Chinese in Shanghai. Through an analysis of marketing strategies they demonstrate how gated developments are billed to represent ideal images of urban living both reflecting and reinforcing the exclusivist housing aspirations and privatist visions of middle-class residents. In some contexts the divisions created by gates demonstrates clear race as well as class divides. For instance, Davis (1992) demonstrates that the creation of walls and gates around suburbs in Los Angeles was entirely for the exclusion of the predominately black and Latino urban poor. Similarly, in South Africa gated communities have been on the rise since residential segregation, put in place under the apartheid regime, has relaxed. Although residential segregation based on poverty, still acutely constructed along race lines, generally remains, the limitations on travelling through wealthy neighbourhoods has relaxed and induced fear of crime among wealthy residents. This has resulted in the explosion of gated developments that are family based and not merely retirement villages. Children growing up in these areas are sheltered from the realities of city living. Participants in Low's study in San Antonio note that families are moving to gated developments to raise children in a protected environment, to escape media attention of child kidnapping and other crimes: 'There are lots of families who have [moved here], in the last couple of years … there's been a lot of "fear flight"' (2001: 53). Issues of race, crime and safety were apparent in Benwell's (2009) work on young people's mobility in Cape Town, South Africa, where his respondents noted that gated communities offered some freedom of movement for young people (albeit within the gates) as lack of access in terms of public transport to these areas meant they were more likely to be visited by those with access to cars. Benwell highlights that the implicit assumptions in these statements were that poor, generally black, people were excluded from frequenting the area without a legitimate reason due to difficult access. Similarly, gated communities restrict young people's ability to move around the city or interact with others of different social and economic status as they have to be taken by parents when they go out of the area. In Montevideo, Uraguay, residents emphasised that children were a key motivator for their move to gated areas (Alvarez-Rivadulla 2007). Although they live segregated lives based on class and income structures in their current neighbourhoods, they felt moving to gated developments increased community interaction among their children and offered them freedom to move around the neighbourhood. In turn, the gated communities become an agent for the socialisation of children in addition to the school and family. As one respondent noted since moving to the community: 'his life "has changed completely". Now he can go on vacation without worrying about robberies. Now his children can play outside without fearing traffic. They can leave their bikes outside without worrying about robberies' (ibid.: 55).

Gated communities, then, are one example by which middle-class families are able to maintain social and economic segregation from other sectors in the city which has implications for children in terms of the protected environment in which they live. They grow up in a neighbourhood which emanates a particular lifestyle associated with the cultural aspect of class where particular values and practices generate and reproduce class divisions. The restricted and protected environments of gated communities are, however, only one example of how middle-class children and youth live in the city. Recently there has been increasing evidence of children and youth being part of the gentrification process in inner-city neighbourhoods.

Gentrification and the inner city

Since the 1960s, there has been a reversal of the trend of the middle classes living out of city centres in large suburban homes. As a product of gentrification processes, young middle-class professionals have moved into restored traditional pre-war housing into contemporary inner-city urban neighbourhoods where residential housing mixes with trendy consumer establishments (see Chapter 3). However, Hall (2007) argues that this has sought to solidify the removal of children in the city as inner-city urban renaissance has resulted in further removal of families from inner-city areas. For example, he argues that one of the key issues for contemporary urban society, at least in the Global North, is the out-migration of affluent families with children from gentrified urban neighbourhoods to more rural or suburban locations, while the café culture and compact planning of inner-city middle-class environments have been designed without really including children. Hall (2007) argues that this is leading to unbalanced population structures in inner-city areas, prompted by the development of gentrified neighbourhoods as part of urban-renaissance schemes. However, recent work has identified two ways in which young people are part of the 'new gentrifying classes': 'family gentrifiers' and 'student gentrifiers'.

Lia Karsten (2003a) demonstrates how children are now also experiencing inner-city living as part of the new class of family gentrifiers. From her work in Amsterdam she argues that gender relations are significant in this process as two-income professional couples reach childbearing/rearing stages in their lives. As both parents are employed, remaining in the city is important for minimising time away from home and reducing the commuting distance between home, work, school and childcare facilities. Under these circumstances inner-city living is seen to have real benefits for children based on access to cultural activities such as visiting museums and galleries, while the ease of meeting with friends in public spaces such as parks was positively valued. Yet, facilities for children are still limited and an increase in these redeveloped areas is called for to meet growing demand from family gentrifiers. A much higher priority to family issues is essential in urban policy and planning (Karsten 2003b). Gary Bridge (2006) supports this notion highlighting that tradeoffs between couples attracting the urban aesthetic of the Victorian or Georgian eras may be offset by the need to access good education and services for their children.

Middle-class young people are not only experiencing inner-city gentrification as part of 'family gentrifiers' but also through 'studentification', which, as Smith (2005) and Ley (1994) argue, is another piece in the jigsaw that sees families pushed from inner-city environments. Since the early 1990s increasing numbers of young people are choosing to carry on with studying and increasing numbers of university places, coupled with the conversion of former technical colleges into 'new' universities, has seen a rise in the student rental market developing in enclaves near campuses in university towns (Smith and Holt 2007). Unlike family gentrifiers, students are seen to be bringing down neighbourhoods, producing what Ley (1996: 181) states as 'emergent youth ghettos nestled symbiotically around inner city university campuses', introducing 'a distinctive sub-culture'. The first part of this quote highlights the (relative) economic poverty of students and the transience of their time in the area for engaging in the redevelopment of the urban fabric, which often remains run down and outdated, yet the latter part of the quote demonstrates their (relatively) privileged position as students and the ability to acquire and sophisticate their cultural capital through the specific enclaves creating their own student identity (Smith and Holt 2007). Students are in fact apprentice gentrifiers using student areas as a gateway to becoming young professional gentrifiers on graduation and securing employment.

The experiences then of non-poor children living in city environments can be viewed in different ways depending on location. These contrasting examples highlight the importance of place for growing up in the city, where the protection of the suburbs and rise in gated development living contrasts significantly with family gentrifiers, where cosmopolitanism and cultural experience are part of inner-city living. However, there are striking similarities related to the cultural and social capital that is reproduced in these environments. Alvarez-Rivadulla (2007) suggests the existence of a global culture of the urban upper service classes. Both groups share similar economic and cultural capital, generally measured by education, relative to the wider urban societies in which they live. What is also apparent is the choice that is available to families and the notion of the priceless child emerges here, with decisions on where to live and the type of environment selected are centred on providing for children's futures.

Living in poor neighbourhoods

Turning our attention to living in poor neighbourhoods, we now explore the experiences of young people growing up in different types of housing tenure and less affluent areas. Chapter 3 explores the negative conditions facing many poor children in terms of sanitation, health and access to services. In this chapter we focus more directly on the physical fabric of poor neighbourhoods and how this directly affects the social relations and experiences of young people growing up there. Globally there is a vast array of housing types for the urban poor, ranging in many countries of the Global North, from publicly provided housing to shared ownership properties and privately rented apartments or homes, often in formerly public housing stock. In other locations slum and squatter housing is host to vast percentages of the urban poor, sometimes located in illegal or marginal locations. Although varying in degree the experiences for young people growing up here can be precarious, as economic and social potential is threatened by lack of access to good quality education or fear of crime and violence. For young people with disabilities, living in slum housing can be particularly problematic. Lack of access to specialist aids and support can mean that they do not have access to even basic needs. Sen and Goldbart (2005) undertook a needs assessment for children with disabilities (both intellectual and physical) in urban slums in Kolkata, India and identified that families had not been taught how to communicate with their children or given support for toilet training as facilities were not available. Children who could not support themselves were often fed lying down with expensive and inappropriate baby food, on which they frequently choked, because of lack of supportive equipment. Disabled children were also found to have no access to education and, although not despised in the community, were socially isolated from other children.

Social exclusion associated with a discourse of fear surrounding many communities in poverty, whether in the UK (Peckham in London) or South Africa (Cape Flats in Cape Town), has other implications for young people. Weller and Bruegel (2009) note that children from Peckham in their study were more likely to stay at home than spend time outside due to fear of violence in the community. One 11-year-old boy states: 'I don't really go ... out anymore. I've witnessed a lot of fights but it's not safe for me to go outside …' (ibid.: 640). Similar examples can be noted in the Cape Flats communities in South Africa where young people would rather leave their neighbourhoods to avoid becoming engaged in gang activities, dealing drugs and carrying weapons for gang leaders. The crime and violence in neighbourhoods exists in conjunction with high unemployment, alcoholism, drug use, single parenthood and other issues that intensify the experience of poverty. However, by drawing on two distinct examples of city living, it is possible to demonstrate the importance of social networks for sustained urban living.

Living in flats

High-rise flats occupy an ambiguous position in contemporary discussions of urban life as they allow for greater population density in areas where space is a premium, while creating space to produce shops, restaurants and other amenities necessary for the production of social life. On the other hand, high-rise flats are often considered to offer low social cohesion among neighbours and are often blamed for society's problems. In the UK and Europe the poor location and poor quality of many high-rise council developments, a throwback to social housing

Figure 4.1 High-rise development in Leipzig, Germany. (Source: Kathrin Hörschelmann)

developments of the 1970s, are located in areas which lack amenities, positioning them as a no-go area. However, in some societies, notably in SE Asia, with Singapore and Hong Kong being particular examples although followed closely by mainland China, population density is such that a significant proportion of the population, including the middle classes reside in flats. Chau *et al.* (2007) note that pressure is so great on land that high-rise estates comprising blocks of flats of up to 60 storeys are being constructed. This has led Appold and Yuen (2007) to suggest that much of the stigma associated with living in flats is more to do with the symbolic meaning that they have become associated with, especially in Europe, than the actual functionality of the space for bringing up children.

Families are increasingly finding themselves living in flats (see Figure 4.1). Yet, flats generally offer three challenges to family life: they are usually smaller than traditional houses creating more crowding and less privacy which can cause tensions in the home among family members; the distance between outside and inside space can create problems when the supervision of children cannot be easily undertaken alongside other chores; and finally the accessibility of public facilities detracts from family time. Appold and Yuen (2007) focused on Singapore to demonstrate that where living in flats is the norm for families (including middle-class families) these challenges can at least in part be overcome. Large spacious flats overcame problems associated with crowding and privacy while the close proximity to services meant children had easy access to a range of extra-curricular activities. Although they could not demonstrate strong social cohesion among neighbours, they also noted that the cultural capital gained from living in close proximity to amenities fed into the future chances of children. This is in stark contrast to Hassan's (1977) work in the same city focusing on a low-income housing estate where residents had been relocated from squatter settlements. He noted that small flats led to family stress and that neighbourhood context was crucially important for the experiences of growing up in flats. This illustrates that it is not tenure type that may be the issue but what it offers and where it is located that is ultimately important.

Migration and renting in the city

As we noted in Chapter 3, migrants form a significant part of the urban poor, particularly in many parts of the Global South, where cities are still rapidly expanding through rural–urban migration. This often means accessing housing that is less secure, including makeshift shacks constructed in emerging squatter areas, building or renting backyard dwellings in more established low-income neighbourhoods or private short-term renting. The associated difficulties faced by families, especially children, in these circumstances are broadly associated with poor social networks, economic difficulties and transiency. For youth who migrate to the cities in search of employment, the problems are exacerbated when work is not found quickly. Box 4.3 further develops these issues by focusing on the private rental sector in Lesotho.

Despite the problematic issues that are part of city migration for young people, research also demonstrates that social networks are adapted from rural settings and often migrants will locate in areas where they already know someone, resulting in distinct enclaves in the city that are inhabited by families from the same villages and ethnic groups. Abu-Lughod (1961) from closely studying migrancy in Cairo notes that migrants usually make first contact with a friend or relative in the city, perhaps lodging there for a few nights and then finding more permanent accommodation in the same area resulting in the formation of enclaves of ex-villagers sharing a common past in the village and a simultaneous adaptation to the city. This is similar to the infiltration processes that occurred in many Latin American cities in the 1980s whereby groups of families and relatives from the same areas would gradually take over a squatter site. These rural connections might lead to the expectation that social relations would be strong among families moving into the cities but, as Englund (2002) notes from work in Malawi,

Figure 4.2 Squatter housing in Ethiopia. (Source: Lorraine van Blerk)

it can take several years for migrants to acquire enough capital to build a house and so renting is a preferred option. Also, as many do not plan to stay in the city permanently, their investment into where they live is less pronounced, resulting in large areas of squatter housing (see Figure 4.2).

However, migrating to low-income accommodation, often rental accommodation in cities, does not necessarily mean that communities lack social cohesion and that the experience will be negative for children and youth. A clear example of this is among the Jewish immigrant community in US and UK cities where ghettoisation was both a product of poverty and the need for maintaining ethnic cultural identity. However, in some cases it was also a strategy of the planning authorities to limit the impact of large-scale immigration through containment of the population in specified areas (Dennis 1997). The experience for children is one of cultural, economic and racial similarities that are highly dependent on place.

The relative social cohesion illustrated here has rapidly changed as immigration in the US has grown. Since the 1980s the immigrant population has steadily increased and is now one of the most ethnically diverse and fastest growing segments of America's child population (Zhou 1997). The disappearance of industrial work in urban areas where racial minorities are concentrated has reduced the demand for low-skilled and semi-skilled workers, increasing poverty among the most disadvantaged of the ethnic minority population. The implications of this for youth are profound. Immigrant children from middle-class backgrounds benefit from financially secure families, good education and safe neighbourhoods, while those growing up in underprivileged neighbourhoods are surrounded by crime, violence, drugs, poor educational and training choices and may be part of single-parent families, who are more likely to live in poverty. This has resulted in resentment among young Americans towards middle-class culture and is expressed in rebellion to authority and rejection of the goals of achievement and upward mobility. This anti-intellectual strand of American youth culture is particularly evident among minority groups as, in underprivileged neighbourhoods, young people meet peers in school and in their leisure time with these values (Zhou 1997).

Box 4.3 **Renting as a problematic experience in Lesotho**

The informal rental sector is expanding across Africa in terms of its position and significance in housing markets; although the type of housing and provision is varied, large proportions of this type of housing stock are at the lower end of the scale. This includes backyard shacks: these are rental dwellings, one room properties attached to larger houses or located amongst similar properties. The place of governments in the provision of rental housing stock varied across counties and historically. However, what is clear is the increase in privately owned rented accommodation.

This research discusses the particular experiences of migrant children living in rented accommodation, around Maseru, the capital city of Lesotho. Since independence in 1966 Maseru has gown at a steady rate although this has largely gone unplanned. By 1980, peri-urban Maseru already accommodated 80 per cent of the city's population. The lack of controlled urban panning means that the peri-urban sprawl is of fairly low density, mixed housing and it is common to find fairly substantial houses mixed with rows of cheap rental accommodation. These privately rented houses are usually called *malaene*, meaning 'lines', because several individual properties, generally consisting of single rooms, are connected to each other in a line (see Figure 4.3).

The experiences of migrant children living in 'lines' are largely represented as negative, which is in part due to the type of tenure they live in and partly due to their migrant status. Both these points are connected. Unfamiliarity with an area and lack of social networks in the immediate neighbourhood, coupled with the transiency of living in relatively unsecure accommodation, demonstrates problematic living experiences for many. Renting was not only a product of migration but often resulted in further transiency as problems of living in lines increased families' chances of moving again.

> Migration is moving from a place you live at and going to another house, because the owner of the previous house is troublesome, so one feels it's better to move to another house and hopefully the owner there won't be troublesome.
>
> (Ansell and van Blerk 2005: 435)

Problems with owners of the lines was a key reason for moving, either because of inability to pay rent, problems with the number of children and accidental damage of property. One boy highlights how this can result in gossip and slander towards tenants forcing them to move on: 'the owner of the lines talks about you behind your back, telling people you are poor and struggle a lot' (ibid.: 435).

Low-income rental communities are demonstrated, in this example, as noisy, unruly spaces characterised by social conflict highlighting the inability of private rental accommodation to act as a solution, at least socially and economically, to housing shortages in African cities.

Source: adapted from Ansell and van Blerk (2005)

Figure 4.3 'The lines': rented housing in Lesotho. (Source: Lorraine van Blerk)

Learning task: living in the city

Think about the different ways in which young people live in the city. Take the city in which you live or study and identify the different types of spaces that make up the city – for example, the city centre, inner-city neighbourhoods and suburban developments. Think about and describe the differences between these areas. How would children and youth experience living in your city? How might their experiences differ with reference to age, gender, class, disability and sexuality?

Now select a contrasting city, perhaps from the Global South. Go through the same process. What are the different areas you might identify? What are they like? How might children and youth experience living in that city? How might their experiences differ with reference to age, gender, class, disability and sexuality? To help you, you might like to visit the UN-HABITAT website (www.unhabitiat.org) and, in particular, look through the 'State of the Urban Youth 2010/2011: Levelling the Playing Field' report which documents the experiences of youth in the following five cities: Rio de Janeiro (Brazil), Mumbai (India), Kingston (Jamaica), Nairobi (Kenya) and Lagos (Nigeria).

Compare and contrast the experiences of young people living in your city and the city you selected from the Global South.

Playing in the city

Play is a very important aspect of the lives of children and youth: although most teenagers would probably refer to their free time in terms of 'hanging out' or socialising than playing per se. In this section we discuss the experiences of 'play' (encompassing hanging out or 'doing nothing' with friends) across various urban spaces and explore how urban space is integral to this aspect of growing up in the city. The section begins by considering children's play in the city before examining how youth make use of urban environments in their leisure time. No attempt has been made to categorically define children and youth by age as the terms are overlapping and the ways in which young people, reaching the child/youth boundary, use the city can be fluid.

Children's play

Living in urban, as opposed to rural, environments offers many opportunities, as well as constraints, for children and youth to engage in play activities. Jones

(2000) discusses the conceptualisation of the rural as the best place to bring up children because of the freedom they have to explore outdoor environments (in contrast to urban environments). However, there are numerous accounts of young people using urban environments in ingenious ways to engage in social play activities. As recently as one or two generations ago it was part of the fabric of urban neighbourhoods for children to be playing in the streets. Colin Ward (1978/1990) expertly details the way in which children use their agency to colonise small spaces and search out places for imaginative play in the derelict spaces of the urban fabric. This extract from his detailed explanations of children at play shows how the urban environment can offer exciting opportunities:

> There was an almost mischievous glee among the children as they took him [an educator] into the scary area of the town, through alleys, ginnels and tunnels, into the district which had no longer an official residence. The council had here spent one-and-a-half day's labour by three men in bricking up the doors and windows of each of the abandoned houses, but they still sustained a population, the inebriates, the junkies and some bewildered homeless people, who along with the bloated dead domestic animals provided the setting for the use the children made of these abandoned streets. It was here they were able to indulge their appetite for either building or destroying. Their human encounters in this sector of the city were those which the educator would have most wished to spare them. But just because the adult users of this no-mans land were unofficial inhabitants too, they were not a threat to the children's determined use of the area as a place of eerie encounters forbidden games and the acting out of destructive passions.
>
> (Ward, 1978/1990: 41)

Other examples demonstrate the use of city streets for games and other forms of play. In Amsterdam parents recall the outdoor play they used to engage in 20–30 years ago:

> As a child in Westsellingwerf one always played outside; you did not really play at home. People recalled building huts, hunting for eggs, roller-skating in the streets, skipping, and spinning tops. In the evenings an occasional game of korfball might take place in the street; occasionally, a car would pass, but you heard it coming from far. Outside there was always plenty to do; there you met the other village children. The children usually stayed close to home until they were five or six, but later they went farther afield. Except for some deep ditches, they were allowed to go everywhere: in the streets, to the woods. Yet they did not move very far from their own village; they did not get as far as Wolvega, for instance.
>
> (Camstra 1997: 28)

However, Ward (1978/1990) and the nostalgic accounts of adults in Camstra's (1997) work tend to underplay the risks and dangers associated with playing in the streets. Others in his study recognise that dangers have increased, at least in terms of the volume of traffic present in urban areas as the following account of playing in Amsterdam neighbourhoods highlights:

> In Haarlem children used to play football in the streets. In the whole Brouwerstraat there was only one car, the coalman's. Now it is full of cars: you have to go to the Leideplein to play football … One might say that children do not really belong in a city. When you live in a city like Amsterdam, you have to accept a low level of participation in outdoor play and a low level of independence. However, children appreciate the same things in a city as adults: the diversity, the excitement, the unexpected. A street may offer at least as much of a challenge as the woods. Why begrudge them their challenge, just because grown-ups prefer parking-space for their cars?
>
> (Camstra 1997: 28, 40)

Although the city offers opportunities for children to experience their environments through play, discourses on children in the city have moved towards the perceived need for greater control and restriction by adults on where children play and what types of play they engage in, as Valentine and McKendrick (1997) and others (see for example, Smith and Barker 2000b; Skelton 2010) note, at least in middle-class contexts. Valentine (1997b: 61) thus states that parents interviewed in her study 'felt they had to be more protective of their own children than their parents had been of them when they were young because children are perceived to be more at risk today'.

How restrictions are placed on children varies depending on the risks. The main risks of being in public spaces in the city, according to Valentine and McKendrick (1997), are related to stranger-danger and traffic problems. Here, gender, sibling order and ethnicity (among other variables) can play a significant part in how much freedom young people are afforded (Karsten 1998). In terms of traffic dangers, Valentine (1997a) notes that girls are more likely to be given spatial freedom because they are asserted as being able to negotiate public space safely due to having greater self-awareness and sexual maturity, whereas boys were perceived as irrational and irresponsible. On the contrary, girls are more likely to be kept close to home because of parental fears over abduction and stranger-danger. Skelton (2010), from her work in Singapore, illustrates that girls on the cusp of their teenage years are heavily restricted in the urban environment for fear of corruption and outrages of modesty (here referring to inappropriate sexual contact including flashing, offensive name calling, sexual suggestion and so on). Karsten (1996) also notes the impact ethnicity can have on restrictions to outdoor environments. Again, using the example of The Netherlands, she notes how Turkish and Moroccan children grow up in larger families than their Dutch peers and may have to stay closer to home to watch younger siblings as part of cultural expectations. Often this falls to the eldest girls in a family.

Given these fears and restrictions, there is an increasing tendency to regulate children's play, resulting in an increase in the creation of designated play spaces in urban neighbourhoods for children such as adventure playgrounds (to offer special spaces for informal play), but also an active move to include children in the design of these spaces (Chawla 1999; Woolley 2007).

This tendency, while informed partly by the desire to create more child-friendly cities, draws on romanticised notions of childhood and their connections to nature and the rural idyll as the perfect place to locate children in the outdoor environment. Interaction with nature is argued to engage children socially, intellectually and emotionally, processes that are important for their development (Kong 2000). However, their predication to play in nature can be met with adultist fears of the dangers associated with unregulated space. Through their research in Singapore, Kong (2000) and Kong *et al.* (1999) note how the urbanised environment is removing children's nature play. Singapore is

Figure 4.4 'Nature Play' in the city, San Francisco, US. (Source: Kathrin Hörschelmann)

Figure 4.5 Community Garden, Berlin, Germany. (Source: Kathrin Hörschelmann)

highly urbanised and the post-independence government has been confronted with a host of problems related to rapid urbanisation, including housing shortages, population growth, high unemployment and lack of proper infrastructure. Although the government put economic and social programmes in place to counteract these negative impacts, the proposals that were initiated necessitated a degree of land-use planning. This resulted in the clearance of some natural areas including large-scale deforestation of tropical habitats and the removal of coral-fringed coasts. However, this now means Singaporeans have little contact with unmanaged greenery which forms the context in which young people grow up. Children are fearful of animals; pets are not part of high-rise living and learning to grow and care for the environment is missing as parents do not have the knowledge of gardening and caring for wildlife. Kong *et al.* (1999) outline how nature has therefore been constructed in the city to satisfy human desire through policies to green the city. Specially designed parks, cultivated roadside verges and managed messicol vegetation now provide a natural balance in the urban environment. This access to constructed nature versus natural nature presents some opportunities, as well as remaining challenges, for children in the city. Specifically designed and planned environments may reduce potential parental fears of the unknown hazards of nature that children may encounter through falling over and hurting themselves as well as providing an outlet to learn about and connect with nature (Kong 2000; Kong *et al.* 1999).

The development of managed adventure playgrounds and parks mean that, through constructed nature in the city, children are being encouraged to engage with the natural at the same time as removing parental fears and anxieties. De Visscher and Bouverne-de Bie (2008) discuss this process in Ghent, where a play web has been constructed that safely connects play nodes (including parks and leisure centres) in the city with a series of safe pathways. A similar system has been set up in Denver, where 'Learning Landscapes' have been introduced as focal points to offer areas for gathering, natural and wild spaces for exploration, spaces for games and exercise as well as creative play elements (Kingston *et al.* 2007). Similarly, Figure 4.4 illustrates an example of nature play in the city of San Francisco. This is one attempt to integrate children in the city by addressing parental fears. Yet the constructed nature of these playgrounds does not introduce young people to the haphazardness of unplanned spaces. In some contexts however, the unplanned spaces of the city are extremely hazardous and the introduction of community gardens, such as that shown in Figure 4.5, can provide opportunities for safe therapeutic play in the city. For example, the development of a children's garden in Guatemala City offered the children living and working on the city's rubbish dump to have access to learning about nature and a safe place to play and explore (Winterbottom 2008), while the sandpit in Figure 4.6 brings the coastal beaches to central Columbia.

Figure 4.6 Sandpit in a Bogota park. (Source: Susan Mains)

Another approach to children playing in the city has been the removal of children's play from informal, and in some cases formal, outdoor environments through the institutionalisation and commodification of play-space where predominantly middle-class children are removed from playing in outdoor public environments and placed in specialised play settings (Karsten 1996; Smith and Barker 2000b; McKendrick *et al.* 2000a, 2000b). Figure 4.7 provides one example of these spaces.

Commodification describes the process whereby objects, places and even relationships or less tangible aspects of life become 'purchasable'. It increasingly affects young people's leisure spaces. Rising household incomes in the Global North have led to both greater diversification and commodification of play. The expansion of commercial leisure spaces has had the beneficial effect of asserting children's right to play in otherwise adult dominated spaces, such as motorway stations or shopping centres. For McKendrick *et al.*, it challenges the prevailing order and has 'asserted children's right to play-space in parts of the built environment which were hitherto perceived as almost exclusively adult domains' (2000a: 101). While concerns have been raised that such spaces increase children's segregation into designated areas while being designed to accommodate the interests of adults (Jones and Cunningham 1999), McKendrick *et al.* (2000b) evaluate them as a positive extension to children's environments. They also find that commercial play-spaces have an important role to play in continuing social traditions such as children's birthday parties at a time when the home has become increasingly private and is less frequently used as a site for socialising. Further, they offer a greater variety of play opportunities than parents may be able to provide themselves in the home.

Despite these benefits, there are restrictions on the use of such facilities related to income, access to transport and ability. Commercial leisure spaces may facilitate the inclusion of some minority children, but they are clearly not environments for *all*. They also only provide spaces for children under adult supervision, where parental presence is a prerequisite. Thus, in exchange for partaking in the exciting play opportunities that commercial outlets offer, children sacrifice a degree of control over their own play. McKendrick *et al.* (2000b) further show that children's involvement in decision making about visits to commercial playgrounds is highly restricted: 'According to parents, only one-third of decisions to visit commercial playgrounds involve children as decision makers; almost one-tenth of cases are a "joint-decision" between adults and children, while just over one-quarter are child-led' (ibid.: 307). Children's leisure interests often matter less than parents' interests in defusing their children's play activities outside the home, providing opportunities for relaxation and letting trained playworkers take responsibility. This leads McKendrick *et al.* (2000b) to conclude that:

> commercial playgrounds are packaging a composite of children's *adventurous* play for children, children's *safe* play for adults and a 'space' in which parents can relax. What at first appears to be a space for children's play is only this in as much as it is contingent upon a series of preconditions about the nature of play that is permitted.
>
> (McKendrick *et al.*: 2000b: 312)

The geography and practices of play are changing, however, not only through commodification but also because of the increasing institutionalisation of after-school leisure time. Smith and Barker (2000a, 2000b) have researched this trend in the British context. They argue that out-of-school clubs, which expanded particularly during the Labour Party era from 1997, impact in significant ways on the social and institutional framework of childhood in Britain, as it is increasingly performed in institutions where adults shape children's use of space and time. Out-of-school clubs are regarded as a place to play by children yet, just like as we saw for commercial playgrounds, they are 'absent from making decisions about the types of play activities on offer' (Smith and Barker 2000a: 248). Children also have little control over the visual appearance of the club space, especially where it has to fit in with other uses, such as in schools. Smith and Barker have found that the boundaries between 'school time' and 'out-of-school

Figure 4.7 Commercial play centre, UK. (Source: Lorraine van Blerk)

time' are increasingly blurred, not just because of spatial continuity, but also because of ongoing supervision by both playworkers and teachers. They further note that the activities of girls and boys are differently structured and perceived. Boys are using outdoor spaces more, while girls play primarily indoors.

Playworkers judge boys' and girls' requests differently, often responding more positively to girls, while boys' behaviour is regarded as more problematic, not least because it tends to exclude girls. The authors also highlight how age and ethnicity affect children's experiences of out-of-school clubs, showing that older children's interests are frequently marginalised and seen as problematic, while ethnic minority children rarely find their identities reflected and embraced in more than tokenistic ways (see Box 4.4).

Box 4.4 **Out-of-school clubs and ethnic diversity**

Space for ethnic diversity?

Research has highlighted the importance of ethnicity to experiences of childhood (James *et al.* 1998) and children's geographies (Valentine 1997b). Ethnicity had a significant impact upon children's experience of place and use of space in the out-of-school clubs taking part in the research ... For ethnic minority children, the out-of-school club was conceptualized primarily as a place to meet and spend time with friends ...

[C]lubs with a significant number of children from ethnic minority communities became important sites for the reproduction of cultural identity. One club, for example, in an area with a large Afro-Caribbean community, had Afro-Caribbean history and culture at the centre of its activities and image ... The children responded positively to the ways in which the club promoted a sense of pride in their Afro-Caribbean identity and began to create their own activities and games to reinforce this. One group of children, for example, routinely played a game to restyle each other's hair in various Afro styles from the last 30 years.

However, most children from ethnic minority groups attended clubs in which the majority of children were white. Moreover, in most of these clubs all the playworkers were also white. In these clubs there was little or no 'place' for the positive promotion of ethnic diversity, evidenced by the lack of posters, toys or books reflecting non-white groups. Indeed ethnic diversity was largely ignored, with the common exception of well-meaning but largely tokenistic attempts at serving food from around the world. These attempts were mostly done without context and children often constructed such activities as 'bizarre'. For children from ethnic minority groups the opportunity to contest such *representations* was limited by the fact that adult playworkers retained control over the food provided, the posters put on walls and the resources bought for the clubs. Moreover, for a minority of children attending such clubs, the feeling that they, as black or Asian children, were 'out of place' in these environments was exacerbated by these processes.

Source: taken from Smith and Barker (2000b: 327–329)

For those who cannot afford such luxuries, the streets often becomes a no-go area, as Weller and Bruegel (2009) note from their work in inner London suburbs, such as Peckham. Here they found children in communities retreated indoors for play and recreation for fear of the dangerous and violent encounters that occurred in their areas. For example, their interviewees mentioned fighting as a regular part of peer interaction, which in some cases has led to knife violence and murder as was the case in the murder of Damilola Taylor in 2000. One boy asserts how his neighbourhood makes it difficult for children to play outside: 'My estate's very tatty and rough … there's lots of drunk people around the flats and there's lots of drug dealing but … We don't really mix in, we keep to ourselves' (Reay and Lucey 2000: 415).

However, under different circumstances streets still offer a place of socialisation and exploration. As Sarah points out: 'Nothing will happen to me because it is so safe around my flats for me. Even though they take drugs and that, it is really safe 'cos they all know me so they wouldn't do anything'(Reay and Lucey 2000: 416). This is nowhere more visible than in many of the slum communities in the cities of the Global South. Abu-Ghazzeh (1998) and Chawla (2002a) note how limited indoor space means that many activities are conducted outside people's homes. Children use the spaces in the streets around their communities to engage in play activities while adults use the streets for cooking, washing, working and socialising. In part, the business of the environment means that younger children can be watched as they play, offering some element of parental responsibility. Informal play constitutes a large part of most slum children's play activities, which observations in Dhaka city attest to:

> Informal play was observed in the streets of the colonies (public housing areas), at the entrance of access roads in the planned residential area and in the alleys of unplanned residential areas. In most cases, children were found

> playing on the paved access roads, narrow streets or uneven grounds close to their houses. There was no maintenance of these informal play areas. Most of these areas were noisy, unclean, unfenced and littered with rubbish.
>
> (Ahmend and Sohail 2008: 266)

Leisure and socialising in the city

Moving on to consider the experiences of youth and their use of the city, a different picture emerges. While younger children's need for play spaces is much acknowledged and accommodated in different ways by providing public and private playgrounds, we have noted that older children and adolescents frequently find their demands for leisure space neglected or regarded as problematic. We have shown in Chapter 3 how welfare spending cuts are affecting public provision particularly for young people from poorer neighbourhoods, while in Chapter 2 we argued that representations of older children and adolescents as dangerous and irresponsible limit their access to public space. Nonetheless, older youth socialise in a variety of settings outside the institutionalised spaces of school and work or the adult-dominated space of the home. The work of Hugh Matthews and colleagues has been enormously significant in detailing the importance of street spaces for young people in the construction of their identities and interactions with peers.

In many ways the street is seen as a third space for young people to express themselves away from the restriction and controlling environments of the home and school (Matthews *et al.* 2000a). The street has cultural importance in that it offers young people the opportunity to assert independence and develop a collective identity. As far back as the 1970s with the emergence of youth subcultures, in part developed out of resistance to dominant middle-class culture, the streets became highly significant for identity formation of the groups of the time: the mods, rockers and skinheads. Brighton witnessed numerous clashes between groups asserting their presence over territories (Hebidge 1979). Today, youth use the urban environment, possibly in less overt ways, but essentially as part of their identity formation and collective sense of belonging. Matthews *et al.* (2000a) highlights, from their work in UK East Midlands towns, that it is often marginal spaces that are used in this way, particularly at night when other street users are out of the streets. Although often asserted as a male domain, Tucker (2003) notes that street spaces are also important for girls both as a place to meet friends and socialise but also as a space for both sexes to meet and interact.

The creation of 'youth only' spaces ensures that identity formation can be achieved away from adults, but at the same time makes adults fearful of youth on the streets and their potential to engage in criminal behaviour (see Young 2003 for a non-UK example). This has resulted in a moral panic about 'dangerous' youth where the presence of young people in public space is considered frightening and a threat to public order. Valentine (2004) states that this has led to popular claims that liberal approaches to children's welfare and children's rights have eroded young people's deference to adults, undermining the subtle regulatory regimes by which adults maintain their hegemony in public space. Over the recent past the public realm, rather than being a social order of civility, sociability and tolerance, has increasingly become one of apprehension and insecurity. Encounters with 'difference' are being read not as pleasurable and part of the vitality of the streets but rather as potentially threatening and dangerous. In this context, young people's nonconformity and disorderly behaviour is often read as a threat to the peace and order of the street.

This means that youth can be inadvertently excluded from many spaces that they use through adult control. As well as the shopping mall (see Chapter 5), the street and urban square offer sites for semi-autonomous interaction with other young people in public space, yet their presence here is often contested and the actions of older youth are seen to conflict with the interests of both adults and younger children. Table 4.1 highlights three key issues that marginalise youths' ability to hang-out in the streets.

Table 4.1 Marginalisation of youth (Malone and Hasluck 1998)

1. The **physical form of the neighbourhood** can often deter young people from occupying places that are away from the main streets and centres – places where they can retreat from the adult gaze and exert their own identity on a place. Badly-lit alleyways and abandoned buildings are often associated with crime and drug-related activity in particular – which many young teenagers fear.
2. The **commercialisation of many youth spaces** has resulted in geographies of exclusion for many teenagers. Revamping and investment in for example, local sports complexes often results in exclusion, as young people cannot afford to pay the entrance fee.
3. Young people growing up in deprived neighbourhoods with high crime rates can have **fears related to their personal safety**, resulting in their inability to freely access their local environments.

The street as a social space in the Global South

The streets, and other public and semi-public environments, are used in a variety of ways by young people growing up in cities around the world. Where youth make up a large proportion of the population and unemployment is high, it is inevitable that young people will form a large proportion of those seeking work. In much of Africa, employment and work are seen as particular virtues for young men and part of the transition to adulthood (see Chapter 3). By contrast, youth hanging out on the main city streets epitomises young people's marginalisation in this context as poor, unemployed and lazy. From her work in Ghana, Langevang (2008) notes that groups of young men on the streets are viewed as signs of idleness and unemployment and not looked upon favourably by the adult population. Despite this, their ability to claim space in the street for socialising and networking through the development of youth clubs is an important step in challenging adult authority. From the inside these meeting spaces were in turn viewed as places for the development of social networks.

By contrast, the back streets, or streets of slum neighbourhoods, in many cities in the Global South are an inherently social space, where as we noted the combining of daily chores outside the home enables greater opportunities for

Figure 4.8 Playing in the streets, Malawi. (Source: Lorraine van Blerk)

Figure 4.9 Playing in the streets, Islamabad. (Source: Lorraine van Blerk)

children to play (see Figures 4.8 and 4.9 for examples of street play). For youth these practices can both offer opportunities and constraints in their ability to use the streets around their homes for meeting friends. In Zambia, Gough (2008) notes that the quantity of activities undertaken in the streets provides opportunities for youth to socialise with their peers while performing daily tasks such as washing clothes and gives them an opportunity to engage in cultural activities at the same time. However, this also causes tensions as living in close proximity to other households can lead to conflict where young people end up in arguments with adults over playing loud music. Similarly in Recife, Brazil, the streets are used in particular ways for youth living in the favelas. Dalsgaard *et al.* (2008) highlight that here most young people do not have their own room and therefore space and social relationships are played out differently for different groups. Where houses are small and streets too narrow for cars, they are essentially meeting spaces for youth. They hang out, talk with friends and engage in informal work activities. Here their lives are embedded in 'high-density social networks' (ibid.: 55). However, for middle-class youth the streets are spaces of threatening social contacts where 'servants, workers and those who try to make a living in the public space (mainly sellers and beggars) walk the unkempt sidewalks, while members of wealthier families prefer to go by car. Better off young people rarely hang out in the streets' (ibid.: 55). In the Global South, then, the street as a social space for youth is class based, increasing the segregation between rich and poor. For wealthier youth little is to be gained by hanging out in the streets, it rarely increases their social capital, while poor youth develop networks for employment and social engagement in the streets.

Other leisure spaces

There are a range of sites, however, that young people use in order to socialise without constant adult supervision. Outdoor environments such as parks and beaches can thus offer important refuges for independent youth interaction (Noack and Silbereisen 1988; Sommer 1990; Owens 1988, 1994; Van Roosmalen and Krahn 1996). In the research project that Kathrin Hörschelmann conducted together with Nadine Schäfer in the East German city of Leipzig (2005, 2007), urban parks and the shores of artificial lakes created along the fringes of the city were favourite meeting places for older young people (see Figure 4.10).

Here, they arranged to see their friends regularly and spent much of their leisure time when the weather allowed it. The many unsupervised beaches allowed them some reprieve from an everyday life that was otherwise highly structured and institutionalised, starting with the journey to school and the school day itself, followed by after-school clubs, organised leisure activities, homework clubs, time spent helping parents or siblings and expectations about mealtimes and family time in the evening. Those who lived in suburban areas further often used the street as a playground or shared private gardens, as illustrated in Figure 4.11. They highlighted that this gave them a strong sense of belonging and allowed

Figure 4.10 Meeting friends by the lakes, Leipzig, Germany. (Source: Anonymous research participant, Leipzig project 2003, Kathrin Hörschelmann)

them to forge friendships outside of school. Yet, sometimes suburban housing estates can be places of isolation for young people, particularly if they are located at a distance from social networks developed through schools and other leisure activities. The experience of young people in Japan's new town's exemplifies that in such instances youth travel into the city centres to access social interaction that is not available in their local communities (see Box 4.5).

Other important sites for after-school entertainment are sports grounds such as football pitches, ice rinks, tennis courts or swimming pools (Van Roosmalen and Krahn 1996; Owens 1994; Korpela 1989; cf. Bain 2003). Young people use these both for exercise and for socialising. It is important, however, to remember that boys and girls often use such sports facilities in very different ways. In Leipzig, many youth clubs sought to offer sports facilities. Hörschelmann and Schäfer (2005, 2007) noticed, however, that these primarily catered for boys' interests and that girls felt excluded from them, partly because participating in 'male' sports was regarded as unfeminine and they felt too observed. In these cases, the sports grounds divided young people by gender rather than providing spaces for interaction and allowed boys to visually and verbally dominate the clubs. Other youth centres sought to counteract this by offering either joint or both girl- and boy-only activities.

Figure 4.11 Playing in the suburban street, Leipzig, Germany. (Source: Anonymous research participant, Leipzig project 2003, Kathrin Hörschelmann)

Box 4.5 **Kaori Momura: young people and Kobe's new towns (original case study)**

Japanese new towns are master-planned communities constructed after the Second World War in response to the housing shortage and population explosion in major cities. Known for its modern urban planning, Kobe City has constructed several of these developments in its northern outskirts, connected to the centre city area and beyond by public transportation systems. Partly modelled after UK new towns, they are essentially bedroom communities that host residences and facilities that cater toward family needs – but in some cases also host research and industrial parks and various institutional buildings. In contemporary Japan, they represent the normalised landscape of ideal family living – a perfect place for child rearing complete with modern amenities derived from Western living styles. Examining young people's lives in urban settings such as new towns reveal not only the processes of marginalisation of young people in public spheres, but how categories of youth identities are defined and negotiated by various agents.

In the current era of a greying Japanese society, early new towns are also experiencing the aging of both residents and the built environment. Older parents are now entering their senior years and their grown children who initially moved to, or were born in, new towns are now no longer the primary users of local schools and playgrounds. Due to a lack of local places associated with youth culture, those young people, old enough to have more spatial and financial freedom, often 'exit' new towns and head to other areas of the city to spend their personal time. Centre city areas of Kobe and Osaka are where young people can experience dynamic urbanism, co-exist with various people, express themselves through consumption practices and participate in cultural activities. In addition, information and communication technologies such as mobile phones have allowed them to interact with each other across geographic space rather than only in designated spots. In some new towns such as *Gakuentoshi* (Academic town) where several high schools, universities, educational institutions, research parks and public schools are located, one might expect the presence of numerous students to add a vibrant atmosphere to the area and to see a youthful cultural scene. Instead, its landscape is generally orderly, clean, quiet, sterile and lifeless. There are heavy regulations on how such spaces can be used, especially in relation to commercial and religious activities. Students, who have to abide by the school regulations both within and beyond school premises, also face many regulations while occupying the rigidly divided functional spaces of living, working, learning, playing and relaxing. As a result, many of the student and young people's activities are 'contained' within school boundaries, student buildings, shopping complexes and domestic spaces.

The commuting hours and after school hours are when the presence of young people is most observed in the town centres – where the subway stations, bus terminals, associated shopping complexes and open squares are located. Many young Japanese are just walking across the squares heading to their homes or to the stations but some 'hang out' at the fast food restaurants or congregate around the square. In the squares, there are warning signs posted by the authorities prohibiting people from a wide range of activities such as riding motorcycles, mopeds, bicycles, skateboards, roller skates and also prohibiting potentially 'dangerous' activities such as ball games. Any activities and behaviours that do not conform – even mundane activities such as walking across certain neighbourhoods – are considered a nuisance, deviant, threatening or transgressive. These are public spaces – but spaces that cater to only a certain 'public.' While there are occasional instances where skateboarders, musicians, dancers, etc. played in the town squares, they are not reoccurring cultural scenes.

The lack of presence of young people and youth culture suggests that the new towns have the tendency to exclude the behaviours that are not wanted by residents. Yet in the process, Japanese youth are marginalised from community place-making in the spaces which they reside and attend educational institutions. This has future implications in terms of an erosion of cultural and social vitality as well as diversity in the communities as their aging processes continue.

The home can also become an important site for carving out space to be alone or with friends. Where young people have a room of their own and the school day ends before parents return from work, or where parents are happy to grant them some autonomy in the home, a young person's room can become a place of solitude and independence as well as a site for expressing identity. It is rarely the case, however, that young people are allowed to, or have the space to, meet with several friends at home. Socialising at home is restricted primarily to close friends, allowing for more intimate relationships, but rarely for the maintenance of extensive friendship networks. Parents do, in some cases, help their children to create semi-autonomous spaces, as Hörschelmann and Schäfer (2005, 2007) witnessed in Leipzig with a friendship group of special needs students. These young people had been allowed to decorate the attic of the house of one of their friends and to use it exclusively for their own purposes. Thus, they had been able to not only occupy a space for their own activities, but also to make it their own by choosing how to decorate it. Since most of the friends lived in small housing estate flats, the attic room was crucial to allow them to socialise after school and became an anchor in their daily leisure time.

Learning task

With a group of friends, fellow students or young people who you can contact through a local school or social club, produce individual mental maps on the places that each of you use most frequently on a day-to-day basis. You may also want to produce maps for different seasons, parts of the week or month, depending on how much your daily activities change over time. As a student, you may want to draw these maps based on your memory of everyday life as a school pupil. In addition to the more 'factual' maps of daily routines and frequently visited places, you may want to draw a map of the places that you feel are most important to you. You could try to represent these places in different sizes and colours, in order to add further depth to your map. As you compare the maps you have drawn and discuss them with the other participants, ask particularly which places seem to occupy centre stage in young people's daily lives, which of these are related to societal expectations and to the institutionalisation of youth, how much freedom you have to travel independently between places, which places are more likely to allow for the spending of unsupervised time with other young people, which places are emotionally important to you and why, and to what extent you feel you belong or do not belong to the different places you have identified. Which aspects and factors make you feel part of a place and which make you feel marginalised or unwelcome?

The night-time economy

Apart from the streets and other leisure spaces, the night-time economy is a further important collection of socialising spaces for youth in the city. Nightclubs, bars, cinemas and other privately owned entertainment venues also occupy an important place in the leisure activities of older adolescents and young adults (Chatterton and Hollands 2002, 2003; Hollands 2002). Being able to visit these venues depends strongly, of course, on a young person's age and their income situation. While some dance venues offer special events for 'teenagers' and many cinemas have afternoon shows, nightclubs and bars are the preserve of over-16 to over-21 year olds, depending on national regulations.

Participating in nightlife entertainment is a key aspect of socialising for older youth and a rite of passage for those just coming of age. It is without a doubt a very popular activity for young people and many cities have sought to respond to this interest by providing a more diverse and extensive entertainment infrastructure targeted at the group of youthful consumers who are earning an income but remain relatively free from family responsibilities until later in life. The extension of youth that has occurred in the Global North as a result of extended education and changing lifestyle options is contributing to this trend. As Chatterton and Holland (2003) explain:

> the development of a prolonged 'post-adolescent' phase and rapidly changing labour market transitions has meant that young people are continuing to engage in youth cultural activity for much longer periods of time … Visiting bars, pubs and clubs is a core element of many young people's lifestyles. In the UK, 80 per cent visited pubs and clubs in 1999, an increase of 12 per cent over the previous five years (Mintel 2000: 15). The 15–24 age group is also ten times more likely than the general population to be a frequent visitor to a club, with 52 per cent going once a month or more (Mintel, 1998: 22) … There has also been a move away from traditional nightclubs, and their association with seediness, violence and excess, in the wake of the phenomenon of 'clubbing', which emerged from the 'one-nation' dance, rave and – to a certain extent – drug cultures of the late 1980s and early 1990s (Collins 1997). While the club scene has diversified, grown and fractured along the lines of a number of smaller consumer groups, since the mid-1990s it has also been commercialised and a distinction between underground and mainstream clubs has been clearly drawn (Thornton 1995). The experience of going to bars and pubs has also been transformed as traditional ale-houses, bars and taverns have given way to the emergence of style and café-bar venues and hybrid bar/clubs.
>
> (Chatterton and Hollands 2003: 68f)

An important aspect of these changes has been a shift in urban governance which has led to large-scale investments by corporate entertainment providers as a key part of urban regeneration. While this investment has been welcome by many for improving and extending leisure provision for older youth in cities, Chatterton and Hollands (2002, 2003; also see Hollands 2002) argue that it has led to the creation of a more sanitised, monotone nightlife mainstream with the effects of increasing surveillance, the enforcement of middle-class cultural values and moral codes, as well as reducing or pushing to the margins possibilities for 'residual' and 'alternative' nightlife:

> [M]ainstream nightlife is a 'normative landscape' in which particular actions and behaviours have become pre-inscribed, tolerated and accepted, while others are not. This is a geography of common sense which renders unacceptable

> the other, the different, the dirty. One has to look a certain way (through designer clothes), be expected to pay certain prices (for designer beers) and accept certain codes and regulations (from door staff).
>
> (Chatterton and Hollands 2003: 88)

Those who either do not wish to comply with the moral codes of these establishments or are excluded from them due to cost or appearance thus have fewer and fewer opportunities for alternative nightlife entertainment. They are also left behind in the race for conspicuous leisure consumption and find themselves in the shadow of the new cities of spectacle, 'hanging out' on street corners, urban squares, parks, car parks or housing estates where their presence gives rise to moral panics about troublesome youth:

> Connected to the industrial city, residual spaces are now surplus to requirements in the newly emerging post-industrial corporate landscape, replete with its themed fantasy world and expanding consumer power. Similarly, residual youth groupings, including the young unemployed, homeless, poor and often those from ethnic backgrounds, are excluded, segregated, incorporated, policed and in some cases literally 'swept off the streets'.
>
> As such, the younger generation of this 'socially excluded' section of the population have a limited number of choices of nightlife. They can steer clear of the increasingly 'middle class' city (Ley 1996) and socialise with what is left of their rapidly declining communities, and in some cases ghettos. Community pubs on estates, social clubs, house parties and the street (Toon 2000) may remain their only choice of venue. Others continue to inhabit those declining traditional city spaces made up of 'rough' market pubs, alehouses, saloons and taverns, earmarked for redevelopment, while local police often view them as sites of petty crime. Finally, some venture out and attempt to consume the 'bottom end' of mainstream commercial provision.
>
> (Chatterton and Hollands 2003: 88)

The landscapes of urban nightlife created by recent bouts of regeneration are thus far more divisive than they seem if we only focus on those using and enjoying mainstream provisions. However, it is possible to opt out of this culture for young people who are essentially excluded and create alternative relaxation spaces in the nighttime economy. Drawing on the work of Valentine *et al.* (2010), Box 4.6 demonstrates how young British Muslims have created an alternative night-time economy.

Box 4.6 **Young Muslims, alcohol and the night-time economy**

Alcohol has become central to much of the night-time activities that youth participate in when socialising and frequenting pubs, clubs and bars around the city. However, where ethnicity, culture, religion or family circumstance mean young people reject this as part of their socialising activities, they can become marginalised in such settings. For young Muslims alcohol is forbidden as part of their religious and cultural values and therefore creates tensions for those that do not wish to be in environments where alcohol is served. As one young Muslim asserts: 'I wouldn't be in an environment where there's alcohol at all ... Not at night, during the day only for shopping, for socialising no, stay away from that place too much corruption' (Valentine *et al.* 2010: 18).

Rather than bending to conformity, Pakistani Muslim young people have created their own oppositional leisure spaces which include small independent businesses, cafes, canteens and other types of social spaces where alcohol is not present yet offers similar opportunities to socialise and meet peers. This enables them to maintain their own faith values and strengthen their religious commitments and identities without missing out on the opportunities created by the night-time economy. In Stoke on Trent, the location of the study, these spaces are generally located in communities, instead of the main city centre, that are not frequented by the white majority. Therefore although such establishments create an outlet for young Pakistani Muslims, they also effectively contribute to the social segregation of the Muslim population from the white majority night-time economy leisure spaces.

Source: taken from Valentine *et al.* (2010)

Sexuality and the night-time economy

As we have seen, then, night-time economy is diversified for young people and can take a variety of forms associated or not with leisure activities of drinking and drug consumption, dancing and socialising in bars and cafes. However, it has particular importance in the lives of young lesbians and gay men. Unlike heterosexual youth who have access to a number of varied spaces and networks for socialising and developing their identities as they approach adulthood, including the normative spaces of school and home, young gay men and lesbians establishing their sexual identities do not have the same opportunities available to them in their local communities. Instead, they have identified the importance of the 'scene' for coming out, where the 'scene' usually represents a distinct area of clubs/bars, commercial spaces and community centres all contained in a particular part of the city (Valentine and Skelton, 2003).

Homosexuality has historically more often been associated with the city, looking back to examples of gay enclaves such as Berlin's Nollendorfplatz in the interwar period which was host to over 40 gay bars (Andersson 2010). Gentrified 'gay ghettos' (Sibalis 2004) or villages that comprise the 'scene' have most significantly emerged over recent decades in many cities in the Global North such as Stonewall in New York, Soho in London, Le Marais in Paris (Sibalis 2004) and more recently in Asia (Aldrich 2004) and South Africa, for example Cape Town's gay village in de Waterkant. It is these areas that are of particular importance in the night-time economy for young gay men and lesbians (Valentine and Skelton 2003). The 'scene' can offer a socialising space for homosexual youth, but also acts as a space for gaining social acceptance and commitment that can take the form of substitute family relationships. In addition, the 'scene' offers a space for experimentation and developing an identity as a young gay man or lesbian is highlighted by the participants in Valentine and Skelton's (2003) study, who talked of trying out different dress codes and behaviours as part of establishing themselves. However, they also note that young people are particularly vulnerable to risks in the 'scene'. Youth, particularly those who may be confused about their sexuality, and perhaps suffering from low self-esteem, can face risks such as pressure into unsafe sex from older members of the clubs and bars, violence and even social pressure to identify prematurely as gay or lesbian. These risks are heightened in those enclaves of the night-time economy that are particularly public displays of sexual encounter. As Andersson (2010) demonstrates that although much of the literature suggests gay villages represent a new homonormative and sanitised gay culture, they co-exist beside those spaces that are overtly sexual and often hedonistic, such as Vauxhall in London where public sex has moved off the streets and into clubs and bars in a new strategy of urban governance.

Other markers of social difference such as disability, ethnicity and religion also intersect with sexuality in the city that result in different reactions to the openness of the 'scene'. In particular, lesbians frequented the bars and clubs of gay villages much less than gay men, highlighting that the 'scene' reacts against stable relationships with girlfriend pinching taking place. Further, various other identities also impact on access to the 'scene'. Valentine and Skelton (2003) note that young D/deaf lesbians and gay men were positive about the scene, yet those identifying as Christian were more critical of it as a socialising space. One participant, Gareth, felt his identifying as Christian was seen as problematic for many lesbians and gay men in the 'scene'. This places the gay 'scene' as offering important social networks for young people in one area of their lives and which plays a dominant part. However, for youth with conflicting identities such as Christian or Muslim religious sympathies, the gay and lesbian night-time economy can also be exclusionary.

Conclusion

This chapter has considered the ways in which young people experience growing up in the city from a variety of perspectives. Markers of social difference, such as gender, age, ethnicity, sexuality, disability and class, have infused through the chapter an understanding of the unique ways in which young people's experiences of living, playing and socialising in the city are diversified. In addition, the physical space of urban environments points to contrasts across young people's lives that range from the type of housing tenure they live in and how this can bring them into close proximity to, or effectively remove them from, experiencing certain spaces. Consider for a moment the children in Lily Kong's work that had little experience of wild nature in their managed and manicured urban environments contrasting with the natural haven offered to children through Daniel Winterbottom's garden in the garbage dump and how these children experienced nature in their environments differently.

In addition, the ways in which young people experience growing up in the city is a relational experience. Families and other institutions play an important role in shaping the ways in which children and youth are enabled and constrained in the city through the ways in which they shape and manage various aspects of living, playing and socialising.

Questions for discussion

1. How does the historical emergence of childhood as a distinct phase from adulthood and its role in the modernisation/urbanisation process of Western capitalist states aid us in our understanding of young people's experiences of growing up in the city?
2. How do markers of social difference, such as gender, age, class/family income, sexuality and disability affect the experiences children and youth have of growing up in the city?
3. In what ways are children and youth conceptualised through their play/leisure in the city?

Suggested reading

Gagen, E. (2006) 'Measuring the soul: psychological technologies and the production of physical health in Progressive Era America', *Environment and Planning D: Society and Space* 24: 827–849.

Karsten, L. (2003a) 'Family gentrifiers: challenging the city as a place simultaneously to build a career and to raise children', *Urban Studies* 40(12): 2573–2584.

Hansen, K. *et al.* (2008) *Youth and the City in the Global South*, Bloomington: Indiana University Press.

Valentine, G., Holloway, S. and Jayne, M. (2010) 'Contemporary cultures of abstinence and the nighttime economy: Muslim attitudes towards alcohol and the implications for social cohesion', *Environment and Planning A* 42: 8–22.

Useful websites

Queer Youth Network: http://www.queeryouth.org.uk/community/

United Nations Youth and disability: http:social.un.org/index/Youth/World ProgrammeofActionforYouth/Disability.aspx

UNICEF Innocenti Research Centre: http://www.unicef-irc.org/

5 Globalisation and youth culture

In this chapter we will:

- explore how the lives of children and youth are influenced by and respond to the intersection of global-local networks in the city
- discuss the effects of commercialisation on the organisation and experience of leisure
- examine the extent to which urban youth subcultures challenge adult dominance in public space and enable creative engagements with the city's materials and landscapes.

Introduction

In Chapters 3 and 4 we considered how structural inequalities affect the life chances of different groups of children and youth in cities around the globe and how young people nonetheless make the city at least partly their own through everyday practices of living, playing and working. The causes of many of the structural issues affecting young people go far beyond the local and national, however (Katz 2004), especially as global divisions of labour have emerged and neoliberal agendas have come to dominate international policy making. In keeping with our aim of showing the extent to which cities not only shape the lives of children and youth, but are also made through the spatial practices of young people, we focus in this chapter on the ways in which children and youth engage with the globalisation of their life worlds (Massey 1995b; Castells 2000; Sassen 2000, 2001; Hopkins 2010), accommodating themselves within as well as appropriating and changing the circumstances of their lives.

The chapter explores the local-global urban worlds of children and young people through examining how their activity spaces are moulded by trans-local forces as well as how their own actions affect the make-up and experience of urban space. We will focus on questions such as the privatisation of leisure, the role of young people as consumers, engagements with global youth cultures and the performativity and practice of urban youth subcultures. The latter show perhaps in the most striking way how young people actively participate in the production of urban space. We will ask to what extent these subcultures can be seen in terms of resistance, appropriation or accommodation.

Before turning to these questions, however, we sketch some of the theoretical frameworks that are particularly helpful for understanding children and young people's positioning in global-local network space. We offer a summary of debates on globalisation, introduce the concept of the 'milieu' and explain to what extent an emphasis on spatial practices is important for understanding children and young people's engagements with the 'glocal' city.

Global cities – local lives?

Urban researchers have long emphasised the role of trans-local flows and networks for the emergence of cities (Massey 1995b, 2005; Sassen 2000, 2001; Amin and Thrift 2002). While debates about globalisation frequently underline the novelty of trans-local connections, cities have always been sites of intersection, or 'nodal points' in the words of Lewis Mumford (1937). Trade, for instance, was a major factor contributing to the growth of cities in Egypt, Mesopotamia and Aztec Mexico, while European medieval cities frequently achieved independence from feudal landlords due to their ability to accumulate wealth through trade, which led to the emergence of finance capital and brought skilled workers into the cities. Colonial cities, on the other hand, emerged as a result of expansionist, imperialist policies and today continue to bear the traces of past periods of 'globalisation'. We therefore need to be careful not to overestimate the novelty of 'globalisation' and focus more in our discussion on the particular types of global-local intersections that characterise cities in different parts of the world today. Theorists of globalisation have emphasised the increasing speed and reach of global transactions, the extension and intensification of worldwide social relations as well as the development of a more globally oriented imaginary, all of which produce a greater sense of global connectedness as well as a heightened sense of global risks (Beck and Beck-Gernsheim 2002; Giddens 1994; Featherstone 1995; Hannerz 1996; Robertson 1992). The processes of 'distanciation' and 'disembedding' that globalisation theorists describe, however, affect not only the global scale, but equally transform the 'local' and indeed complicate the distinction between scales, as it becomes difficult to distinguish between local and global causes, effects and relations. This has led the theorist Roland Robertson (1992) to advocate the concept of 'glocalisation' instead.

The fortunes of cities depend increasingly on their location within global-local flows. 'Global cities' become the major nodes for flows of finance, information, transport, resources and people. They are, however, not only characterised by the mobility of high-flying elites and spaces for the management of transnational flows. 'Global cities' also depend on the availability of cheap labour, often provided by immigrants, and are dotted with pockets of deprivation, stagnation and immobility, which necessitates a more careful examination of globalisation's proclaimed disembedding effects (Kaplan 1996; Bauman 1998). Further, while globalisation is often represented as a universal process happening *to* people, we need to consider how the 'global' is created through social actions and as such is structured by power relations and differently engaged with by distinct social actors. Such an approach is particularly important if we seek to understand the different ways in which children and young people encounter and engage with globalising processes in urban space. Their spatial worlds today extend far beyond the strictly 'local', partly because of the deterritorialising processes that they experience in their everyday lives (Tomlinson 1999; Skelton and Allen 1999) and partly because, in addition to travel and migration, they are likely to engage with trans-local worlds through media and other cultural flows (Skelton and Valentine 1998; Appadurai 1996). 'Locality' thus has less salience for individuals and social relations than assumed in earlier humanist geographies of place and belonging, though a sense of place and belonging continues to be important for many young people.

Box 5.1 **Deterritorialisation and reterritorialisation**

Social and cultural aspects of globalisation are often described in terms of deterritorialisation and reterritorialisation. Deterritorialisation refers to the weakening of ties between culture and place, which can occur both as a result of increasing engagement with cultural practices from other places and societies and as a consequence of the movement of people in and out of place, whether voluntarily or forced. In a 'mobile', globalising world, cultural ideas and practices transcend fixed territorial boundaries, leading to continuous processes of disembedding and re-embedding that are variously experienced as unsettling, uprooting, enriching, imposed and/or liberating. As communities engage with cultural ideas and practices from and in other places, these ideas and practices are appropriated and partially adapted in a processes of reterritorialisation. Cultures are simultaneously deterritorialised and reterritorialised in a world 'on the move'. As they transgress former boundaries of place, they gain new meanings and are incorporated and adapted in the places they have moved to.

These broader claims not withstanding, however, we need to be careful not to assume that children and young people experience the de- and re-territorialising processes associated with globalisation in the same way (Hörschelmann and Schäfer 2005, 2007; Hörschelmann 2009; see Box 5.1). As Fincher and Jacobs (1998: 1) have argued, '[w]e inhabit different cities even from those inhabited by our most immediate neighbours'. Urban dwellers negotiate variable positionings in urban society and the disadvantages and privileges that are made and remade by economic, social and cultural institutions contribute to producing a far from level playing field (ibid). In their research on youth cultures in post-socialist Russia, Pilkington *et al.* (2002) thus found significant differences between young people's access to and modes of engagement with global cultural flows:

> For those young people whose horizons stretched only as far as their home district, globalization had meant an increasingly narrow focus on locale. Heightened competition for places in higher education, for professional training that might guarantee a 'real' salary, and above all, the desire for a degree of social security, meant that young people's lives often focused intensively on study, work, and earning money (legally or illegally).
>
> (Pilkington 2002: 164)

For many young people in Russia, 'globalisation' and its corollary of neoliberal, post-socialist transition has thus meant increased insecurity and competition, leading to a focus on 'localised' strategies rather than the cosmopolitan, playful engagements with global youth cultures often highlighted in literature on postmodern youth. Pilkington *et al.* (2002) nonetheless underline the significance of recognising all of these young people's agency in consuming and producing 'glocal' cultures through everyday practice: 'It is in these daily, concrete, and often mundane practices, it is suggested, that young people define distinctive "strategies" for contemporary "glocal" living' (Pilkington with Starkova 2002: 101). The authors define 'cultural practice as a series of local engagements with global cultural forms' (ibid: 102). In the absence of sufficient publicly funded, 'alternative' youth leisure spaces, young people's leisure practices are focused on 'getting together' with peers at home, on the street, in the countryside, or in the yard, sites through which affective microgroups are constituted that lend them a sense of belonging and identity not necessarily based on consumption (Maffesoli 1996).

The milieu

A useful concept for understanding the intersections of local-global scales in young people's lives is suggested by Dürrschmidt (1997) in his review of a research project on globalisation and everyday life in the city conducted in the London district of Wandsworth (Eade 1997). Dürrschmidt adopts the concept of the milieu to describe how we make 'the' world into 'our' world. Based on the phenomenological philosophy of Albrecht Schütz (1972), he argues that individuals develop familiar understandings of the world and habitual ways of acting in the context of their everyday lives. These habitual ways of thinking and acting, developed in a specific milieu, give us implicit knowledge of how to deal with others and how to respond to changing circumstances (also see Connolly 1998; Bourdieu 1977, 1984). Individuals perceive and seek to organise their surroundings in self-determined and familiar ways, thus making 'the' world into 'their' world. The concept of the milieu is not the same as that of locale, as the milieu is shaped by wider social structures and increasingly by global relations and flows. Yet, it enables us to retain a sense of the locatedness of young people's lives, whose embeddedness in particular structures and relations informs how they engage with and, in their own way, *make* a changing world their own. The case study below (Box 5.2) illustrates this in a particularly telling way through the example of Ramesh, who has experienced the rapid globalisation of Kathamndu, Napal, while building an imaginary, yet highly meaningful relation to New York. This imaginary relation alters both his own identity and how he experiences everyday life in Kathmandu.

Box 5.2 **Living in the global city**

Kathmandu is a city undergoing a speedy initiation into the late twentieth-century world of international aid, global trade, mass tourism and electronic mass media. Before 1951 public education and all communications (travel, trade, books, cinema, etc.) between Kathmandu and 'the outside' were either banned or the exclusive privilege of local elites. Yet in the 1990s Kathmandu has become one of South Asia's busier air transportation hubs hosting close to a quarter of a million tourists a year; fashionable shops are jammed with imported electronic consumer goods making Kathmandu 'the Hong Kong of South Asia'; televisions and VCRs have become standard features of urban middle-class homes [...]

I had not seen Ramesh for over a year when I glimpsed him out of the corner of my eye while riding my bike down a crowded, narrow Kathmandu street one chilly spring morning in 1991. Ramesh looked considerably more gaunt and bedraggled than I remembered him from our earlier acquaintance when he was a regular at a drug-free youth centre set up for recovering addicts. I had heard from others that Ramesh had relapsed into his heroin habit and the jittery but probing look in his eyes when I hailed him made me think twice about the wisdom of re-establishing this relationship.

At 21, Ramesh was living life close to the edge, though seven or eight years earlier no one would have predicted the rough times that lay ahead. Ramesh's parents had moved to Kathmandu from an Eastern hill district when he was in his early teens. He had attended a respected English medium school in the valley and learned to speak competent, colloquial English. He had first tried heroin as a high school student but over the course of a few years, during which his mother died and his father married a woman with several sons, Ramesh developed a habit which grew out of control ... By early 1991 he had been in and out of drug rehabilitation seven times and had little more than the clothes on his back and a few rupees in his pocket. He lived by his wits day to day, hustling tourists, taking profits on petty commodity transactions and running a variety of scams such as sewing foreign labels into locally produced garments [...]

Ramesh was a special connoisseur of films, books, magazine articles – anything he could find – having to do with America, especially New York City. He knew all the city's boroughs and landmarks and was especially intrigued by 'the Bronx', a place he brought up again and again in our conversations. From dozens of tough-guy films and gangster or mafia novels, Ramesh had constructed a detailed image of a New York City street culture full of drugs, thugs and gangs. He frequently compared Kathmandu's street life with that of New York, as when he explained how Kathmandu street gangs take 'tabs' (specific combinations of prescription drug tablets) before going to a fight, 'just like in the Bronx' [...]

Ironically, it seemed sometimes as though Ramesh already lived in New York. 'The Bronx' in particular seemed to be a kind of shadow universe where his mind roamed while his body navigated the streets of Kathmandu. 'The Bronx' – with its street-smarts and anti-heroic codes of valour – was often the standard of reality against which he measured his own existence. For Ramesh, 'the Bronx' seemed to offer a way of understanding his own life, a life that he hated, yet which he could link with a way of existence at the modern metropole. Ramesh's vision of 'the Bronx' allowed him to identify his own existence as at least some version of modernity, even if it lacked the late night bars, video games and a host of other modern accroutements that he had never seen in more than two dimensions.

Source: Liechty (1995: 166–201)

Consumption and the commodification of leisure

One of the ways in which young people interact with and appropriate the 'global' is through consumption. This includes not only their use and enjoyment of global cultural products, such as media and music, but also daily encounters in the globalising city as a site of spectacle and consumption. The place of consumption in young people's lives has received much critical attention in recent years. On the one hand, consumption is seen as a practice that allows the flexible adoption, expression and modification of lifestyles. It has been argued to be more important for the formation of identities in the twenty-first century than a young person's position in the labour market (Miles 2000; Bennett 2000). Consumption, in this understanding, opens up opportunities for experimentation and for choosing identities, which allows young people a degree of independence and choice that they otherwise rarely experience in adult-dominated society. This understanding of consumption as an opportunity for young people is related to the 'discovery' of youthful consumers in the second half of the twentieth century:

> The model of the teenager that arose in the early twentieth century gained cultural significance in the post-World War II economy of growth and affluence (Bennett 1999), where young, middle-class consumers were freed from wage-earner responsibilities. As a result, the teenage identity became inextricably linked to leisure and hedonic consumption.
>
> During this post-World War II phase, young people came to be seen as a lucrative market segment and the very embodiment of the emerging mass culture (Morin 1962). This cultural viewpoint led to the marketing industry becoming increasingly preoccupied with youth, as exemplified by marketing books such as The Teenage Consumer by Mark Abrams (1959). Abram's report has often been referred to as the first evidence of the conspicuous consumption habits of young consumers and, as such, marks the beginning of seeing youth as a segment and hence as a distinct marketing identity.
>
> (Kjeldgaard and Askegaard 2006: 232)

Contemporary Western understandings of youth thus have much to do with the marketing of youthfulness and the discovery of young people as consumers. This has, on the other hand, led to critiques of young people's participation in conspicuous consumption, where their activities are seen as problematic expressions of hedonism and selfishness. Young people here are seen as easily misled by the attractions of mass-produced commodities, as unrestrained and in need of adult stewardship (Griffin 1997; Kjeldgaard and Askegaard 2006; Zukin and Smith Maguire 2004). They are, indeed, increasingly targeted as consumers at a younger and younger age, particularly in advertising for media products and computer technology. Fears over the alleged malleability of young people's desires and minds repeatedly evoke moral panics that recall the conventional construction of Dionysian versus Apollonian childhood (see Chapter 2, Griffin 1997).

Figure 5.1 Young shoppers, UK. (Source: Lorraine van Blerk)

Geographies of consumption

Geographical and sociological research on young people's consumption practices in the city has highlighted a number of both opportunities and constraints. It has cautioned against simplistic understandings of young people as either active or passive consumers, emphasising the different ways in which young people engage with and gain access to consumer opportunities and the negotiations that are involved in carving out a consumption space for young people. Consumption has been examined primarily as an aspect of young people's leisure time, although there are other areas of daily life, such as education, that are increasingly commodified. We have already discussed two types of leisure spaces that are increasingly commodified, the playground and the night-time economy (see Chapter 4). Here, we consider how the shopping mall, as a particular site of consumption, is appropriated by young people for a range of social activities.

Research by Matthews *et al.* (2000b) and Vanderbeck and Johnson (2000) has shown how shopping malls can become important sites for socialisation that allow young people a degree of autonomy away from the adult gaze. The mall 'acts as a convenient social theatre providing an important social venue for young people' (Matthews *et al.* 2000b: 268). It 'is used and imagined as an alternative to both home and streets, providing a combination of safety and security with possibilities for socialisation and entertainment absent in either space' (Vanderbeck and Johnson 2000: 20). Both Matthews *et al.* (2000b) and Vanderbeck and Johnson (2000) underline the sense of safety that young people experience in the shopping mall compared with other locations for meeting and socialising, such as the open street, parks or urban squares. Yet, their presence in the mall rarely goes uncontested. While young people are targeted and welcome as consumers, their use of the mall as a place to 'hang out' and meet friends contradicts the functional logic of a place designed to facilitate shopping. As Matthews *et al.* (2000b: 288) explain,

> when in the shopping mall young people are never out of sight. Malls are places of scrutiny and the geography of the mall, with its open spaces, bright lights and busy walkways assures visibility and exposes their presence … When young people gather in shopping malls their presence is often deemed unacceptable, partly because these places are commonly interpreted to be an extension of the public realm of adults (in which young people have no place when adults are around) and partly because their behaviour is perceived to be at odds with the norm. The hiring of private security guards and the use of surveillance cameras ensures a moral regulation of public space that controls both access and behaviour and those who do not belong are moved on …

Contrary to images of conspicuous consumption, Matthews *et al.*'s and Vanderbeck and Johnson's research also highlights the financial constraints which limit many young people's participation in consumer culture, producing a sense of marginalisation even as they spend much of their leisure time in the midst of consumer society: 'For young people with limited resources malls are often contradictory places. On the one hand, they represent the images and consumption to which many young people aspire; on the other, much of what is available is out of reach and underlines their marginality within society' (Matthews *et al.* 2000b: 288). Research by Elizabeth Chin (2001) with working-class African-American families in the north-east of the United States has likewise contradicted the image of African-American teenagers as avid consumers. Low household incomes, the scarcity of small stores in segregated neighbourhoods, and the suspicion aroused by black teenagers frequenting suburban malls severely restricted their purchasing power and opportunities for using shopping malls. Their shopping was focused mainly on practical goods (cf. Zukin and Smith Maguire 2004). We thus need to be careful not to overestimate the spending power of young people and their independence as consumers.

Consumption is not a free for all. Young people engage with it creatively and certainly use it to represent particular identities, yet their lifestyle choices are circumscribed by family income, parental influence, socio-cultural values and gender ideologies. It is thus a little too early, and certainly highly ethnocentric and elitist, to argue that consumption has replaced other anchors of identity (Miles 2000; Chatterton and Hollands 2002, 2003).

Where young people engage actively in consumption through frequenting urban shopping centres and malls, their activities also continue to be surveilled and regarded by adults with suspicion, as Matthews *et al.* (2000b) show in their research of a UK shopping mall (Box 5.3). The activities of youthful consumers are tightly monitored and acceptance into the space of the mall is dependent upon conforming to adult defined norms of consumption behaviour.

The shopping mall is also a useful example to demonstrate the tensions between youth and other city users in their ability to use space. As a semi-public space, it offers youth a collective meeting point, a cultural place where they can engage with global cultures and consumer identities. It is also a relatively safe place for enjoying social life given the semi-private regulation that takes place through CCTV surveillance and security patrols (see Vanderbeck and Johnson 2000). Similarly, but also uniquely, Skelton (2010) notes that youth are welcomed in the Singaporean shopping malls both by their parents, because it offers them safety, and private business, as it offers an lead in to consumer culture. In other contexts though the mall as a shared space with adults demonstrates that tensions also exist there. One boy exerts: ‘See that security guard, he picks on us. What a prat! We don’t go though. If he says get going or something, then we smooch around a bit until we find somewhere else where he can’t see us. We won’t leave until we want to go’ (Vanderbeck and Johnson, 2000).

Box 5.3 **The unacceptable flâneur: the shopping mall as a teenage hangout**

Young people use shopping malls as a social theatre for interacting with other young people. They value them for providing a warm, dry place where they can stand or sit and meet up with their friends. The mall is seen as a place of excitement, where special things happen:

> I like it here [outside HMV music shop] ... we can listen to the music, chat with each other ... see our friends ... two or three of us will turn up here and soon there'll be five maybe more of us ... we know that we will see someone we know. (Girl aged 15, Grosvenor Centre, Northampton)

The shopping mall becomes a setting of social inclusion for these young people, 'a convenient and accessible meeting point where they can gather to assert their sense of belonging and group membership'.

While outdoor spaces afford young people an opportunity to create a (semi-) autonomous space or the 'stage' where they attempt to play out their social life while maximising their informal control (Matthews *et al.* 2000b), however, shopping malls are places of scrutiny and the geography of the mall, with its open spaces, bright lights and busy walkways assures visibility and exposes the presence of young people.

As adults attempt to define and regulate the mall for their own purposes, collisions and conflicts with young people are commonplace. They are often asked to move on by security guards, shopkeepers, the police and vigilant adults. This happens particularly in places that are perceived to be for movement and flux, even when there is enough space for small groups to gather:

> We were moved the other day ... just chatting and that, looking over the balcony over there ... we were told that we can't stand there as we were blocking the way, yet there was plenty of space for people to pass. (Girl aged 15, Grosvenor Centre)
>
> I'd only just turned up ... I saw me mates sitting around outside HMV ... went across and this security geezer comes up and tells us we got to go ... we weren't doing nothing ... having a fag ... we says why? He says just get going. (Boy aged 15, Grosvenor Centre) [...]

Young people use a range of strategies when faced with the demand to move on, such as moving to a different part of the mall or leaving, but returning after a short period. Only a small minority feel sufficiently intimidated not to return. It is unclear to many young people why they are being targeted in this way. They find their exclusion from these spaces unjust and insist on their right to use them. The young people interviewed by Matthews *et al.* also express a strong sense of wanting to belong. Their suggestions for how to change a shopping mall include providing more places where young people can hang out without fear of being moved on and encouraging shopkeepers to be more tolerant of their presence. Only few suggest removing or reducing the number of security guards. The majority see moral regulation and control as an acceptable part of that society to which they seek to belong.

Source: Matthews *et al.* (2000b: 285–291)

Learning task

In order to complete this task, you should ideally make contact with a group of young shoppers who are willing to be accompanied during a visit to a shopping centre. If possible, you could choose two different locations, such as a mall and an outdoor market. You could start by conducting some observations of the interactions between groups of youth and other users of the space in general, and then focus on the experiences of your chosen group in more detail. You could conduct interviews with the young people to get a better understanding of their experiences, asking especially to what extent they encounter restrictions, stereotypes and negative attitudes towards them as users of shopping malls or outdoor markets, whether they have ever been moved on and why, what they enjoy most about the experience of shopping, to what extent they also shop for others, e.g. family members, as part of their responsibilities for them, and in what ways their shopping practices are affected by issues such as limited access to finance, gender identities, peer pressures and the desire for branded goods. Read the paper by Matthews *et al.* on young shoppers as 'unacceptable flâneurs' (2000b, Box 5.3) before completing this task.

Urban youth cultures – between resistance and leisure

Another important way in which young people carve out social space in the city is through participation in subcultural activities. In Chapter 2, we discussed the association of youth with deviance. This is a definition we find again in descriptions of urban youth subcultures, which are frequently seen in negative terms as a deviation from socially accepted norms. Researchers of youth culture have, however, challenged this negative connotation to argue that the subcultures associated with young people are an expression of their agency and creativity as well as a response to adult hegemony (Hebidge 1979; Clarke *et al.* 1976; Willis 1977; McRobbie 1993).

We need to be careful, however, not to over-romanticise the potential and intention of subcultural groups to resist dominant social norms. Not only are they often tightly interwoven with and display structures similar to the wider social order (Gelder 1997), their intention may also be less to challenge 'mainstream' society than to carve out a niche for a different way of enacting the self, the social and space. Focusing only on oppositional aspects of subculture further runs the risk of overlooking internal stratifications, such as hierarchies between group members, between groups and between subcultures. Thornton found this to be an important characteristic of 'club cultures', where young people compete over status and draw distinctions between members depending on their possession of what she calls 'subcultural capital'. This type of capital is displayed by wearing the right clothes, listening to the most cutting-edge music, being 'in-the-know', visiting particular clubs, etc. (Thornton 1995). The term 'subcultural capital' invokes Pierre Bourdieu's (1977, 1984) notion of social, cultural and symbolic capital which is acquired in particular social fields and structured along lines of social and cultural status. These forms of capital are embodied in the 'habitus' of individuals, consisting amongst other things of particular ways of acting and speaking, clothing, tastes in food, music, art, home decorations, etc., as well as of who one knows and is recognised by. Youth subcultures frequently display hierarchies built on similar forms of distinction and thus the lines between them and so-called 'mainstream society' are more fluid than is sometimes assumed. We will demonstrate this further in the two case studies below, where we examine the embodied spatial and cultural practices of skateboarders and graffiti writers.

Finally, through commodification, youth subcultures are often co-opted by the market and arguably lose much of their oppositional potential as they become a sellable good. Subcultures often veer between the desire to remain 'authentic' and the pressures as well as the attractions of a market that offers opportunities for financial gain and social recognition.

In focusing on youth subcultures, we wish to highlight the ways in which they carve out spaces for living differently in the city. What we do not wish to imply, however, is that they represent 'youth' in some straightforward way. It is a cliché to equate subculture with youth. Not only is the age spectrum of members wider than that usually associated with youth, but they are also frequently dominated by young men, thus telling us little about girls' cultural interests, activities and creativities (McRobbie 1993; Tucker 2003), while being actively engaged in by only a minority of young people. Few commit in an exclusive way to a subculture or acquire the skills and status of a fully integrated member of the 'scene'. It is much more usual for young people to move between subcultural trends and to only adopt aspects of their fashion and musical tastes, often depending on preferences in their peer group (Miles 2000).

Understanding urban youth subcultures

Much work on youth subcultures has focused on aspects of style and meaning (Clarke *et al.* 1976; Hebidge 1979). Their particular spatial and embodied practices have received only a little attention by contrast. Yet, urban youth subcultures such as skateboarding, free walking or graffiti engage creatively with place and perform their identities through it. They frequently establish connections to specific locations through which a sense of belonging emerges, while not being necessarily defined by localness (Gelder 1997). Urban youth subcultures resemble a bricolage of global-local influences, where translocal cultural flows are negotiated and performed in ways that often explicitly work through the city's urban materials (Hörschelmann and Schäfer 2005). The two examples which we describe in the following sections show in different ways how urban youth subcultures work with and recreate the materialities of the city, sometimes only to produce transient landscapes, other times to establish more permanent boundaries that mark out separate places as in gang graffiti.

In order to understand the spatial practices of these subcultures, it is helpful to draw on the theoretical work of Michel de Certeau and Henry Lefebvre (see Box 5.4), both of whom emphasise the significance of practice in the making of space. This is highly relevant for the two subcultures which we discuss in the following sections: skateboarding and graffiti. Skateboarding is an embodied activity that, through physical skill, appropriates urban spaces in ways that are contradictory to their explicit design and commonplace use. It traces myriad trajectories in the urban landscape and claims temporary presence without ever belonging or marking out a permanent territory. Graffiti, likewise, is always on the move and, while marking symbolic boundaries in the case of gang graffiti, otherwise makes only temporary claims on an expansive, unbounded urban space. Both skateboarding and graffiti are practised in the interstices of a capitalist city that is organised around private property and regulated by state and private surveillance. Neither is able to claim material ownership of space, though instances of spatial fixing exist with the examples of skateparks or officially sanctioned 'halls of fame'. Both skateboarding as well as graffiti discover different urban spaces, often finding value in marginal sites that have been disregarded by developers and left behind in the race for profit, while appropriating urban space for alternative uses and experiencing it differently.

Box 5.4 **Spatial practices, strategies and tactics in the work of Henry Lefebvre and Michel de Certeau**

Henry Lefebvre's work emphasises practical ways of experiencing and *appropriating* urban space (Borden 2001; Shields 1999; Elden 2004; Elden *et al.* 2003; Kofman and Lebas 1996). He distinguishes between material spatial practices, representations of space and spaces of representation (1991). Urban space is the result of all these three dimensions, including dominant visions of the city, such as in urban planning (representations of space), utopian and personal imaginations of the city (spaces of representation) and the interactions of physical and social flows (material spatial practices). Lefebvre emphasises the role of the body in spatial practices and looks towards the so-called 'spaces of representation' as sites where people reappropriate space in ways that are contradictory to the dominant logic of the capitalist city. In the words of Rob Shields,

> Lefebvre differentiates the popular 'appropriation' of space from the 'dominated' space of the nation-state, or of the capitalist city. The latter is the site of the hegemonic forces of capital, the former the site of possible emergent spatial revolutions. The local and punctual 'détournement' (re-adaptation, hijacking) of space, as in the tradition of occupying key spatial sites or buildings as a means of protest, is similarly an example of the seizure and re-functioning of hegemonic space.
>
> (Shields 1999: 165)

De Certeau distinguishes between two types of practice: strategies and tactics. He describes strategies as

> the calculation (or manipulation) of power relationships that becomes possible as soon as a subject with will and power (a business, an army, a city, a scientific institution) can be isolated. It postulates a *place* that can be delimited as its *own* and serve as the base from which relations with an *exteriority* composed of targets and threats ... can be managed. As in management, every 'strategic' rationalisation seeks first of all to distinguish its 'own' place, that is, the place of its own power and will, from an 'environment'.
>
> (1984: 35f, italics in the original)

Strategies seek to establish and secure 'proper' places that mark, in de Certeau's terms, 'a triumph of place over time' (1984: 36). They produce bounded, owned places which allow the accumulation of resources and the exercise of power towards an 'outside'. A tactic, by contrast, is defined by its inability to establish an own 'proper' place. It always operates within the space of others:

> The place of a tactic belongs to the other. A tactic insinuates itself into the other's place, fragmentarily, without taking it over in its entirety, without being able to keep it at a distance. It has at its disposal no base where it can capitalize on its advantages, prepare its expansions, and secure independence with respect to circumstances. The 'proper' is a victory of space over time. On the contrary, because it does not have a place, a tactic depends on time – it is always on the watch for opportunities that must be seized 'on the wing'. Whatever it wins, it does not keep. It must constantly manipulate events in order to turn them into 'opportunities'. The weak must continually turn to their own ends forces alien to them.
>
> (Ibid.: xix)

Youth subcultures can be seen as encompassing both strategies and tactics. On the one hand, they insinuate themselves tactically on the terrain of a city over which they have little control or power. They appropriate the spaces owned by others in fleeting, temporary ways that transgress boundaries without being able to, or willing to, establish new 'proper' places. On the other hand, however, some youth subcultures edge out areas that they control and draw symbolic as well as material boundaries around their 'turf', as is particularly evident in gang graffiti. Their strategic manoeuvres are targeted at other youth subcultural groups and create zones of inclusion and exclusion.

Skateboarding

Youth subcultures such as skateboarding show explicitly that the city offers up material and imagined spaces for alternative cultural practices. It is too simplistic to regard urban spaces as always and only restrictive to young people. Although urban youth subcultures frequently respond to exclusions from urban space, they adopt a range of both tactics and strategies to claim presence in the city. Skateboarding can be described as a spatial practice that appropriates and reinvents the materials of the city through the deployments of bodily skill. Borden (2001: 12) sees skateboarding as 'a practice, a particular patterning of space-time produced from a specific body-centred origin'. It demands the development over time of embodied techniques for moving across asphalt, railings, curbs, stairs, street furniture and other materials offered up by urban space using only one's body and a skateboard (see Figures 5.2 and 5.3).

Figure 5.2 Skateboarder in Leipzig, Germany. (Source: Anonymous research participant, Leipzig project 2003, Kathrin Hörschelmann)

Figure 5.3 Skateboarder in Karlskrona, Sweden. (Source: James Glendinning)

Skaters explore both central urban spaces, such as shopping centres, sidewalks and plazas, and those 'forgotten' spaces left behind by capitalist urban development, since the most skate-able places are frequently those least built-up and commercialised. In central urban spaces, they more often than not become a target for exclusionary security measures, yet their transit through or temporary use of such spaces reclaims central areas of the city momentarily for otherwise excluded youths and signals possibilities for alternative experiences and uses of urban space beyond the demands of commerce and the capitalist economy. Skateboarding introduces a different rhythm into urban life, while discovering 'empty', 'forgotten' spaces, such as car parks out of hours, side streets, brownfield sites and derelict highways that offer few commercial attractions, but are full of potential for the skater, as expressed by Tom Lock, a contributor to the skateboarding journal 'Document':

> I love those motorway-esque spots that undeniably conjure up the words 'crack' and 'grime'. It's bizarre to think to the eyes of the average Joe these areas appear depressing and down there with the lowest forms of degradation, whereas skaters are the only people that find real joy in these barren, practically ghettoised areas.
>
> (Lock 2005: 71)

Skateboarders distinguish themselves as a subculture by seeking ways of living outside of society, although they are simultaneously in its very heart. Their identity is often built on a sense of being more authentic, more honest and closer to urban living than mainstream society:

> Skateboarders view the world differently from non-skateboarders. An expanse of concrete is just that to someone who has never skated, but to a skateboarder it presents endless possibilities.
>
> (Brooke 1999: 38)

> It was always a real rebel sport, we were always looked down on because of our aggressiveness and our attitude. It became commercially acceptable in the late '70s and '80s because anything and everything skateboarding was so popular … If you don't skate, then you can't relate, and that's all there is to it … The thing that I enjoyed most was skating natural terrain like I still do today. That's like banks, pools and pipes. Stuff that's built by man to serve one purpose, but when you have the mentality of a skater and you see this terrain, you can see moves that these people that built it never saw. Basically, that's your canvas, as for what skating is to me, it's all art form. It's a release of all your energy out on the terrain, your canvas and creation of your own art.
>
> (Tony Alva, in Brooke 1999: 79)

Skateboarding requires significant skill that is developed over long periods of practice. In this sense, the subculture can be seen as producing its own hierarchies based on accomplishment and embodied mastery of technique and terrain, as well as an activity that is structured like a career. The level of skill is often an indication of the dedication of an individual, which in turn leads to the establishment of hierarchies between members of a group (Borden 2001).

> The skater becomes one with his board, while the board in turn translates the language of the terrain … It's a combination of balance, technique, power, knowledge, love, hate, respect and fear – instinctive perception gained only through repeated action. Skateboarding is a moment of glory, of achievement, of unsung triumph. For the man, skateboarding is freedom and youth rediscovered; for the boy, a means of self-expression vital to his being. For you see, skateboarding is the blood, rolling through his veins.
>
> (David Hackett, in Brooke 1999: 176f)

For its most dedicated practitioners, skateboarding becomes not only who they are, but also what they think and dream about even when they do not practise it:

> As a teenager, skateboarding was a passion, an all-consuming activity. When I was not on that plank with four wheels, I was thinking about it, often dreaming about it.
>
> (Tannis Watson, in Brooke 1999: 76)

> Skateboarding to me means freedom, an outlet for any sort of stress and responsibilities. It's my way of expressing myself.
>
> (Tony Hawk, in Brooke 1999: 106)

> Whenever I just cruise down an empty road or sidewalk, doing nothing but feeling the wind in my face, I realise that this is why I do it … It's a feeling of total freedom and perfection which is the goal of life.
>
> ('Lillis', in Brooke 1999: 141)

Although participation in the subculture of skateboarding lends members a sense of distinction from mainstream society, of personal identity and of authenticity, there are clear similarities with other social structures. Skateboarders develop ambition and self-discipline in the interest of self-improvement, and there is a commercial side to skateboarding that includes specialist outlets, fashion trends and the use of skateboarding as a signifier of youth and rebellion in popular music. Skateboarding is also a predominantly male activity and the terms in which it is described by practitioners and to outsiders are often masculinist in tone, emphasising the 'guts', stamina and physical prowess it requires. This is not to say that women have no interest in the sport or are less talented in it, but their presence is rarely highlighted or noted.

Graffiti

Graffiti has become almost synonymous with contemporary Western cities. It often heralds arrival into the city, as train tracks, the underground and underpasses along highways and major roads are some of the most popular sites for graffiti. Though mainly seen as a modern phenomenon, the practice of 'wall writing' has a long history, emerging at least as early as the first cave drawings. In its current expression, however, graffiti emerged in the low-income black and Hispanic inner-city neighbourhoods of New York's South Bronx in the 1970s (Rahn 2002). It was and still is part of the hip-hop subculture that also includes breakdancing, rapping and DJ-ing. Hip-hop graffiti is thus part of a wider subculture that gives visibility to disenfranchised young people as well as a sense of hope and identity. It connects those who can read and write its signs and produces a sense of belonging that is both territorially bound to particular neighbourhood areas and makes claims to spaces owned by others, insinuating itself tactically on urban space. Youth gangs frequently use graffiti to mark out territory and to compete over the control of neighbourhoods:

> Whether in alleys, around hangouts, parks, on main drags, or on major streets, gang members write graffiti for themselves and others who understand their messages as acts of representing. Gang members use graffiti to organize the spaces in which they reside; in the process, they cover the streets with manifestations of their own identity. Like silent sentinels, the graffiti keep watch and inform onlookers of the affiliations and activities that define a certain area.
>
> (Phillips 1999: 135)

While its symbolism and the sheer presence of graffiti is often seen as a direct affront to the values and rights of mainstream society, gang graffiti is primarily a medium for enforcing relationships between gangs. It may express a sense of alienation, discontentment, marginality, repression, resentment and rebellion, yet gangs primarily fight inwards and their rebellion is rarely directed at society at large.

It is important at this point to distinguish between gang graffiti that marks out belonging to and exclusion from certain neighbourhoods and other forms of hip-hop and political graffiti. Political graffiti is explicitly directed at 'wider' society and the state and concerned with expressing opposition to particular issues. In the city of Berlin, for instance, much political graffiti speaks out against gentrification and the privatisation of housing, while some is directed against militarism and imperialism, cuts in welfare spending and fascism. Neo-Nazis, on the other hand, also use the city's walls to promote their anti-immigration politics. Such graffiti uses lettering that is legible to the uninitiated, is usually unadorned and brings across a particular political message. Hip-hop graffiti, on the other hand, is most characteristic for its invention and reinvention of symbolism and painting styles that are illegible to the wider public and that attain meaning only within the context of the hip-hop community. It has its own aesthetic and much competition between hip-hop graffiti writers is based on aesthetic differences and the appraisal of skills, competence and daring. Such graffiti is by definition impermanent. It will fade with time, be painted over or crossed out by competing writers, or be erased by the actions of police or property owners. Like a palimpsest, graffiti is written, erased and written over repeatedly, especially at sites that are symbolic or have become so within the graffiti community.

De Certeau's concept of tactics applies well to hip-hop graffiti, since it seeks no capital ownership of property, moves across the city rather than becoming fixed to one locality, lives within and across the boundaries of the 'proper', is transient, spontaneous, fleeting, ephemeral (Silva 1986) and yet visible, constantly reproduced, marking a visible presence in otherwise exclusionary spaces. Graffiti combines both absence and presence. It brings people into contact who would otherwise rarely interact (Phillips 1999). As it edges itself onto city walls, it keeps alive questions over the ownership of space and reminds us of the different ways in which space can become appropriated – over and beyond the uses intended by planners, architects, builders and owners. The spaces thus appropriated do not always conform with the logic of valuable real estate. Vacant spaces along the margins of the city, such as derelict industrial sites, disused housing, train tracks and underpasses can all become a canvas through which graffiti writers make urban space their own (Rahn 2002). For GENE, a graffiti artist interviewed by Rahn (2002: 168), writing brings 'a dead space to life', while for SHANA, it presents an opportunity to 'just make the city more of my own place' (ibid.). Like her, DSTRBO makes explicit the importance of graffiti for creating a sense of belonging and ownership: 'I just like to see my art up in my neighbourhood' (ibid.):

> Both transients and vandals take the throw-away places of the city and make them their own … These are the places the city has forsaken. No one cares for them. This creates space for the linked elements of transience and vandalism to arrive slowly and take over physically and decoratively … The city's homeless, taggers, and, in well-established areas of abandonment, accomplished hip-hop writers come together, setting their imaginations alight as well as sometimes the buildings they occupy. They make such places their own, not by claiming them as territory but by using the space towards their own end.
>
> (Phillips 1999: 309–10)

Graffiti raises important questions about rights to representation in public space. A sanitised city without graffiti may give an impression of order and 'cleanliness', but it hides contradictory claims over space and creates silences instead of public engagements with difference and the effects of social exclusion. While graffiti is seen by many as vandalism and is an illegal act that damages property, it highlights both the arbitrariness of a middle-class urban aesthetic and some of the exclusions produced in capitalist, adultist space. Graffiti writers have their own sense of aesthetic, with many nuances and differences in style. They challenge the understanding that only blank city walls are 'beautiful' and seek the same right as advertisers and other producers of urban representations to express themselves visibly in urban space. In highlighting thus the potential of graffiti writers to challenge social and aesthetic norms and to bring visibility to those excluded from equal participation in the creation of public space, we need to none the less remain vigilant against over-romanticising the 'rebellious' potential and character of graffiti. Not only does it compete with other interests that may deserve equal recognition and respect, graffiti rarely 'attacks' society explicitly (Phillips 1999) and, like skateboarding, it displays many characteristics that compare with middle-class social norms (Macdonald 2001).

Although graffiti writers often describe what they do in aggressive terms, talking about 'battles', 'bombing', 'killing', etc., these expressions refer to manifestations in graffiti alone and are not to be taken literally. The main goal of hip-hop writers is to represent themselves within the hip-hop writer arena, to establish a name and position, to create rather than to destroy (Phillips 1999). None the less, graffiti writers relish their outsider position and in addition to competing with other gangs or crews, draw much motivation from confrontations with the police and city authorities. The judicial system may thus in fact unwittingly encourage graffiti (Rahn 2002; Austin 2001):

> [w]riters piece and tag as much to achieve the excitement, the 'adrenaline rush' of illicit creativity, as to leave lasting marks or images. Graffiti writing occurs, then, in a context which challenges, defies, and even celebrates the illegality of the act – a context which can only be exacerbated by the harsh efforts of anti-graffiti campaigners.
>
> (Ferrell 1996: 148)

City authorities often react vehemently and with much combative rhetoric to the emergence and spread of graffiti, which is seen as a challenge to property rights as well as an expression of moral and social decay (see Figure 5.4). In Britain, it has become synonymous with antisocial behaviour, while in North America, urban politicians have called on 'moral entrepreneurs' to take a zero-tolerance stance on graffiti. Such campaigns, while producing a moral panic about unruly youths, have not only been limited in effect, they have also set graffiti writers a challenge which they respond to and which adds to their sense of achievement and outlaw identity. At the same time, Cresswell (1992) and Phillips (1999) have argued that graffiti is used as a platform for politicians to seem actively involved and to create an illusion of control.

Graffiti is usually discussed and described by outsiders. There are few research accounts that take seriously the intentions and motivations of graffiti writers themselves and make room for their own explanations. One exception to this is the research conducted by Nancy Macdonald (2001), which follows an ethnographic approach. Macdonald offers an intriguing insight into the life-worlds of graffiti writers and uses this material to challenge accounts of youth subculture that focus on its symbolism at the expense of practice and on contrasting the subculture with mainstream society at the expense of appreciating its own logic and internal dynamic: 'Graffiti is neither mindless nor senseless. Far from it. The time, energy and effort that are put in reflect an underlying rationale and serve a clear and coherent purpose' (2001: 92). She underlines the personal gains, passions and feelings that make graffiti an all-encompassing way of life for many writers: 'It becomes like your lifestyle, you know what I mean, it's a full time thing' (SAEB, in Macdonald 2001: 67).

Figure 5.4 Graffiti in Montpellier, France. (Source: Kathrin Hörschelmann)

Macdonald is careful not to overestimate the rebellious character of graffiti and finds that its social structure is more comparable to middle-class society than commonly assumed. Not only are many of its practitioners in European cities from middle-class families, they also adhere to a similar ethic of ambition, hard work and achievement, which means that graffiti writers develop careers in the scene marked by stages of skills acquisition, visibility and respect. Middle-class values such as dedication and diligence to achieve status are embraced rather than rejected and as such, graffiti is less an act of spite than a practice focused on its own goals and ambitions. While appearing transient, it follows long-term aims and objectives, such as increasing one's skill and reputation by painting bigger, more elaborate pieces in spaces that are difficult to access and/or over a wider terrain. A writer's status will often depend on and develop with their degree of daring, perseverance and accomplishment, ranging from simple 'tags' that can quickly be spread over a wide area but require less artistic skill, to full-colour, large pieces in difficult spots that require not just artistry, but also bodily skill. Graffiti is thus not only a practice of representation, but also a form of material, embodied engagement with the city. This aspect is rarely recognised in research on youth subcultures and deserves greater attention.

Returning to the question of social norms and identities in the graffiti subculture, Macdonald also highlights the masculine overtones of the scene. Most of its performers are male and graffiti is described primarily as 'men's work'. Its internal logic of competition, confrontation and achievement further mark it out as a masculine activity, where reputation is gained by acting tough and daring. Militaristic metaphors are used not only within the scene, but also to describe the 'battle' of graffiti writers with police and city authorities. The respect and recognition of peers further validates a writer's masculinity. This does not preclude women from participation, and indeed the secretive and invisible aspects of the practice of writing graffiti make gender-identification tricky, but the scene leaves little space for expressions of identity and status other than those of hegemonic masculinity.

While for urban geographers and researchers of childhood and youth, researching a subculture such as graffiti is valuable for identifying ways in which young people can contest their exclusion from public space, it is important to remember that such research will necessarily fail to represent the subculture fully and in its own terms. We need to ask whether that should indeed be the aim of research, especially when a gap in understanding preserves important space for the subculture to continue to challenge the status quo. Graffiti writers thrive on being 'misunderstood' and this is exactly what makes their practice challenging. As ZAKI, quoted in Macdonald, explains:

> It's quite a wonderful feeling to be part of something that is misunderstood by the rest of society … I'm glad they don't know, it's something that they'll never understand and if they did understand, would you really want them to in the first place?
>
> (Macdonald 2001: 154)

The authenticity of graffiti is already drawn into doubt by its gradual commercialisation and the use of graffiti in popular culture (Rahn 2002; Macdonald 2001; see Figure 5.5). As researchers, we may wish to resist the urge of making the 'other' entirely knowable and purchasable, for:

> As graffiti culture becomes increasingly packaged as a lifestyle, it becomes sanitized of any political awareness and sold as a consumer good. Youth buy into representations of hip-hop because it suggests danger and transgression, without learning the codes and beliefs driving the community.
>
> (Rahn 2002: 177)

> You can't buy things like that and you can't buy walls either.
>
> (ZAKI, in Macdonald 2001: 172)

Figure 5.5 Sanctioned graffiti: commissioned wall painting in Leipzig, Germany. (Source: Kathrin Hörschelmann)

Graffiti writers do not merely consume, but synthesise and translate their environment into another medium or context (Rahn 2002). As researchers and students we may wish to look at our own practice in a similar way and, with Austin (2001: 271), ask the following visionary question: 'What would a wildly decorated city look like? To which the overriding response must be: what kind of city do people want to live in?'

Urban youth culture in non-Western cities

Much of the literature on youth lifestyles focuses on Western contexts, especially the UK and the US. Both youth culture and consumption continue to be treated in youth research as the primary domain of 'postmodern' Western youths. While on the one hand there has been insufficient attention paid to continuing socio-economic inequalities that affect consumer behaviour and participation in the West (McRobbie 1997), on the other hand, non-Western youths' engagements with globalisation, postmodern lifestyles and cultural hybridity have featured little in cultural and sociological research (Saldanha 2002; also see Skelton and Allen 1999). This scarcity of research might lead to the mistaken assumption that leisure and play are the preserves of wealthy Western children and youths, while elsewhere young people's lives are characterised by tradition, labour and hardship. Although anthropological and geographical research has shown that in many poorer countries, the boundaries between children's play and work are fluid, this does not mean that children do not build play into their daily routines (Punch 2000; Ansell 2005; Katz 2004) or that older youths do not engage creatively with global cultural flows (Farrer 1999, 2002; Saldanha 2002; Pilkington *et al.* 2002).

The two case studies we have selected from South Africa and Indonesia (Boxes 5.4 and 5.5) illustrate not only the creativity with which young people in these countries perform identities (Hammett 2010; Beazley 2002), they also

Figure 5.6 Amusement park, South Africa. (Source: Daryl van Blerk)

demonstrate that young people from diverse socio-economic backgrounds participate in global youth culture and appropriate it in their everyday practices (cf. Beazley 2003, 2008; Beazley and Chakraborty 2008; Butcher and Velayutham 2009; Massey 1998; Nilan 2006; see Figure 5.6). Hammett, in relation to Cape Town hip-hop, demonstrates how global musical influences are reworked in local cultural productions that explicitly focus on tackling inequality, poverty and marginalisation in a particular South African urban context. Beazley, by contrast, shows how street girls in Indonesia use their bodies to reject traditional Indonesian constructions of femininity, i.e. by copying male dress codes, wearing short hair, smoking or taking drugs. Her research addresses a major gap in the otherwise predominantly male-focused literature on youth subcultures (McRobbie 1993, 1997), by showing that girls create their own cultural practices, perform alternative embodied identities and utilise the public space of the street to do so, thus also altering the meaning of public space.

Box 5.4 **Cape Town hip-hop**

Hammett's paper explores Cape Town hip-hop as a site of resilience, reworking and resistance of globalising pressures. It shows how Cape Town hip-hop evolved in a context of contradictory socio-economic and political developments, where marked socio-economic progress since the 1990s has been accompanied by continuing challenges of poverty, HIV/AIDS, unemployment, violent masculinities and crime. Hammett shows how Cape Town hip-hop reworks international influences on youth culture, music and language through engagements with local political issues and concerns:

> Cape Town is host to a vibrant and varied music Scene [...] There is a burgeoning of local hip-hop and rap musicians (primarily from the former 'coloured' group areas). The development of Cape Town's hip-hop scene is rooted in local production and consumption practices [and] in spatial histories of contact and commodification in which local and international influences were reworked in local meanings, dispositions and identities. These artists engage with the social and political context of the city, seeking to address challenges relating to gender, race, poverty, development, social justice, economic inequality, ethnicity, class and social consciousness. The spatial context is significant in the quest for authenticity in hip-hop (Krims 2002). The experience of urbanisation, forced removals and residential restrictions generated specific social and political geographies (see Jansen 2008; Wilkinson 2000). [...] The development of local hip-hop in this context demonstrates hybridisation and reworking of meanings. American

gangsta rap is often posited as a foil to local, critical hip-hop's presentation as a vehicle for social activism rooted in struggle histories (Pritchard 2009). Despite the poverty and social dislocation of the Cape Flats, local hip-hop does not reify consumerism, violence and masculinity in ways associated with American gangsta rap born of similar circumstance (cf. Krims 2002). Instead, Cape Town hip-hop draws upon histories of dispossession and oppression to articulate critical engagements with contemporary social, spatial and economic inequalities. [...]

The socio-spatial context of the Cape Flats, the former 'coloured' group areas of Cape Town, provides a geographical foundation for local hip-hop's politicisation.[...] Historic demographic factors that frame the concentration of 'coloured' communities in these areas are also important to the specific development of the music scene in Cape Town. Thus, hip-hop continues to enjoy a particular appeal within local (coloured) communities, while the influence of kwaito and Afropop remains relatively limited due to the unusual demographic composition of the region. Black Noise formed in 1990, emerging as one of Cape Town's foremost politicised hip-hop groups. Their music continues to address local development, social justice, economic inequality and questions of race and identity. They are involved with social outreach projects including workshops, school tours and fundraising for young hip-hop artists. Emile YX, one of the founding members of the group, whose solo productions critically engage with capitalism, history, social justice, identity and materialism, remains the political driving force in the group. He is heavily involved in social outreach projects, including the Heal the Hood project (a social project aimed at bringing together youth to develop artistic skills, tackle xenophobia and developing social responsibility) and annual hip-hop indaba (a national competition for DJs, MCs and hip-hop groups, the winners of which are funded to compete in the world hip-hop championships). Godessa formed in 2000, releasing their debut single, Social Ills, in 2003 and album, Spillage, in 2004. An all-female three-piece group, Godessa's performances were geared towards providing positive, black female role models and commentating on key social issues including HIV/AIDS, gender violence and media representations of (South) Africa. They worked with social projects including the HIVHOP project (with Bush Radio and Voice of America) to promote AIDS awareness, exchange and awareness raising projects funded by the Netherlands Institute for Southern Africa, and used hip-hop and spoken word in prisons and rehabilitation centres. [...]

Hiphop artists rework local cultural and mediascapes to raise awareness of inequitable conditions while attempting to mobilise resistance and practices of redress. Cape Town hip-hop is not simply about resistance, nor is resistance understood through a simple binary of local/global or hegemonic/subaltern, but demonstrates a more nuanced and symbolic set of encounters. Artists use music and media to rework and question the role of culture in perpetuating socio-economic inequalities and materialism. These practices seek to re-appropriate media for critical engagements within the constraints of commercial and economic imperatives.

Source: Hammett (2010: no page)

Box 5.5 **Girls' street subcultures in Indonesia**

Young people on the streets of Indonesia's cities are marginalised by the state and mainstream ideology, and this marginality is reflected in the spaces they occupy. State Ibusim, represses women to a subordinate position in society as part of the 'New Order' state in Indonesia and therefore girls are not socialised to work in the public area and given domestic roles around the home from an early age. Therefore, street girls are particularly marginalised, seen to be committing a social violation as their presence on the streets contradicts this state ideological discourse on family values where the street is no place for children, especially girls, who are firmly located in the domestic sphere. Just by being on the streets they are seen to be violating constructions of femininity and this positions them at the bottom of the sub-cultural hierarchy, pushed to the margins of street kid sub-culture. Street boys themselves label the girls *rendan*, vagrants wearing make-up, a term the girls find offensive, as it likens the girls to prostitutes placing them as powerless objects of sexist discourse and inferior to others.

Yet, the girls on the street are not merely appendages to street boys' subculture but rather use strategies of resistance to subvert their position, displaying their own subcultural values through psychological, social and spatial practices. As resistance to the alienation they experience on the streets, the girls have carved out their own spaces in the city, most particularly the *Taman* – the city park, where they can retreat to when they have had enough of the male-dominated spaces of the city. It is here they can be themselves and not be subordinate to the street boys when occupying other streets in the city.

> The Taman was a gathering place for a group of about 12 street girls ... These girls slept and kept their possessions in a small house at the gates of the park and, during the day, they would 'hang out' in a sheltered part

> of the building and in other parts of the park. The girls who lived in this park called themselves the 'Park Kids' (Anak Taman). When they were in this space, their behaviour was very different from when they were in the street boys spaces, as they were far more vocal, gregarious and confident. (p. 1675)

The street girls also note how male aggression commands respect on the streets and the home and they emulate it, gaining confidence and status on the streets. In response to their social marginality they adopt styles and behaviours that respond to their position of sexual subordination and is a clear rejection of the Indonesian constructions of femininity. They use their bodies as sites of empowerment, copying male dress codes and markings; wearing jeans and shirts, cutting their hair short, smoking, taking drugs.

> Like the street boys, most of the girls had tattoos, and many of them had the name of their boyfriend tattooed on their hand. As well as tattoos and body piercing, the girls have numerous razor cuts, often in rows up the insides of their arms. These cuts are a sign of their subculture and present a tough and anti-feminine image. Almost all the girls had these scarifications on their arms and I read them as a 'social inscription', which can be understood as a 'public collective ... a sign of belonging to their group' (p. 1676).

Street girls therefore actively subvert their position as subordinate on the street through various acts of resistance displayed as subcultural style. They are 'participating in the spectacle of street kid sub-culture, by 'working a reputation' for themselves' (p. 1680) that gives them respect and a position on the streets.

Source: Beazley (2002: 1665–1684)

Technology and socialising

Carving out space for unsupervised interaction with other young people does not always require the use and occupation of a material space. As social networking sites such as Facebook have shown in particular, modern technology offers numerous original ways for the production of social spaces that can be created literally 'on the move' and on both a momentary and continuous basis. Research by Ito and Okabe (2005) on Japanese young people's mobile phone use illustrates this in the most interesting way. They argue that Japanese young people have less private space than their European or American counterparts, since homes are smaller than in many Western cities and their daily lives are structured around movement between the home and school, making the street a key site for socialising. Mobile phones have 'revolutionized the power-geometry of space–time compression for teens in the home, enabling teens to communicate without the surveillance of parents and siblings' (ibid.: 7):

> youth mobile messaging has worked to construct alternative kinds of intimate 'places' or settings where youth can be in touch with their close peer group or 'full-time intimate community' (Nakajima *et al.* 1999) ... [Y]outh messaging can undermine certain adult-defined prior definitions of social situation and place, but also construct new technosocial situations and new boundaries of identity and place. To say that mobile phones universally cross boundaries, heighten accessibility, and fragment social life is to see only one side of the dynamic social reconfigurations heralded by mobile communications. Mobile phones create new kinds of bounded places that merge the infrastructures of geography and technology, as well as technosocial practices that merge technical standards and social norms.
>
> (Ibid.: 5; see Box 5.6)

The mobile phone allows young people to create a social sphere in-between and within spaces that are dominated by adults and which otherwise provide them with little privacy. Phone use connects with and is influenced by location in urban space, but it also transgresses material and social boundaries, thus enabling new forms of social interaction.

Box 5.6 **Japanese young people's mobile phone use**

The Japanese urban home is tiny by middle-class American standards, and teens and children generally share a room with a sibling or a parent. Most college students in Tokyo live with their parents, often even after they begin work, as the costs of renting an apartment in an urban area are prohibitively high. Because of these factors, Japanese youth generally take to the street to socialize. For high school students, this usually means a stop at a local fast food restaurant on the way home from school. College kids have more time and mobility, gathering in cafés, stores, bars, and karaoke spots. Unlike the US, there is no practice for teens to get their own landline at a certain age, or to have a private phone in their room [...]

Phone ringing is ... considered a violation in public space. [By contrast], mobile e-mail is considered ideal for use in public spaces ... Unlike voice calls, which are generally point-to-point and engrossing, messaging can be a way of maintaining ongoing background awareness of others, and of keeping multiple channels of communication open [...]

While the scale of social relationships and content of communication appears to be similar to what other studies have found in other forms of mediated communication ... the portable format of the mobile phone affords certain distinctive patterns. Heavy mobile email users generally expect those in the intimate circle to be available for communication unless they are sleeping or working. Text messages can be returned discreetly during class, on public transport, or in restaurants, all contexts where voice communication would be inappropriate. Many of the messages that we saw exchanged between this close peer group or between couples included messages that informants described as 'insignificant' or 'not urgent.' [...]

These messages define a social setting that is substantially different from direct interpersonal interaction characteristic of a voice call, text chat, or face-to-face one-on-one interaction. These messages are predicated on the sense of ambient accessibility, a shared virtual space that is generally available between a few friends or with a loved one. They do not require a deliberate 'opening' of a channel of communication, but are based on the expectation that someone is in 'earshot.' From a technology perspective, this differs from PC-based communication because the social expectation is to be almost always connected. This is also not a 'persistent' space as with an online virtual world that exists independent of specific people logging in (Mynatt *et al.* 1997). As a technological system, however, people experience a sense of persistent social space constituted through the periodic exchange of text messages. These messages define a space of peripheral background awareness that is midway between direct interaction and non-interaction.

While mobile phones have become a vehicle for youths to challenge the power-geometries of places such as the home, the classroom, and the street, they have also created new disciplines and power-geometries, the need to be continuously available to friends and lovers, and the need to always carry a functioning mobile device. These disciplines are accompanied by new sets of social expectations and manners. When unable to return a message right away, young people feel that a social expectation has been violated.

In line with the moral panics over *kogyaru* street cultures [media label for street-savvy high school girls], public discourse has associated pagers and mobile phones with bad manners, declining morals, and a low-achievement, pleasure-seeking mentality. Mobile phones continue to be iconic of a fast and footloose street culture beyond the surveillance of the institutions of home and school [...]

Social and cultural research paints a different picture of young people's media adoption. The young people in our studies were highly conscious of mobile phone manners and used their phones to keep in touch almost exclusively with family and close friends from school ... [T]he regulatory and protective functions of institutions such as family and school still dominate the lives of Japanese youth into their twenties.

Source: Ito and Okabe (2005: 127–145)

Conclusion

In this chapter, we have focused on young people's engagements with urban space through leisure and the participation in subcultural activities. We have suggested that youth cultures are an important medium through which children and youth encounter and creatively (re)create the 'global' via their embedded, localised spatial practices. The city as a space of global-local flows can have unmooring effects on young people's identities, yet also opens up opportunities for creativity and for participation in urban life. We have also shown, however, how fears *of* and *for* young people lead to further restrictions on their access to urban public space. Leisure is increasingly linked to consumption for young people: urban regeneration schemes have in many Western, and increasingly non-Western cities created leisure infrastructures for older youth that target the 'hip', 'cool' and 'well-to-do', while alienating and marginalising those who either resent the trend towards greater commercialisation or simply do not have the means to participate in conspicuous leisure and entertainment consumption (Chatterton and Hollands 2002, 2003; Hollands 2002, also see Chapter 4).

The final sections of the chapter have focused more closely on ways in which some groups of young people resist their exclusion from urban public space through participation in subcultures such as skateboarding and graffiti. We have sought to illustrate their potential to contest the hierarchical structuring of capitalist urban space and to creatively reinvent the city through different embodied, spatial practices, while also retaining awareness of the contradictions that arise from the commercialisation of youth subculture and from their own social structuring, which limit the extent to which they challenge and disrupt the 'mainstream'. Such critiques notwithstanding, the practices of subcultural groups raise important questions about legitimised and outlawed forms of living in the city that are deeply related to the issues of citizenship and participation, to which we turn in the next chapter. They challenge us, in particular, to rethink whose spatial activities are accepted as 'normal' and why. In order to answer this question, we need to consider what a *city of difference*, a *radically open* city would look like (Malone 2002).

Questions for discussion

1. To what extent can children and youth be regarded as participants in the production of global spaces in and through the city?
2. How would greater responsiveness to the cultural needs and interests of *diverse* groups of young people transform the city?
3. To what extent do urban youth subcultures challenge the exclusion of young people from public space?

Suggested reading

Borden, I. (2001) *Skateboarding, Space and the City. Architecture and the City*, Oxford: Berg.

Farrer, J. (2002) *Opening Up: Youth Sex Culture and Market Reform in Shanghai*, Chicago: University of Chicago Press.

Gelder, K. and Thornton, S. (eds) (1997) *The Subcultures Reader*, London: Routledge.

Macdonald, D. (2001) *The Graffiti Subculture. Youth, Masculinity and Identity in London and New York*, Basingstoke: Palgrave.

Skelton, T. and Valentine, G. (1998) *Cool Places: Geographies of Youth Cultures*, London: Routledge.

Useful websites

UNICEF Voices of Youth: http://www.voicesofyouth.org/en

UNFPA – youth page – check out in particular the publications on young people and culture and growing up urban: http://www.unfpa.org/public/home/adolescents

6 Participation and active citizenship in the city

In this chapter we will:

- outline the main arguments for promoting young people's participation in urban planning
- discuss a range of participatory methods
- consider ways of enabling more effective participation and of avoiding the pitfalls of tokenism, decoration and manipulation
- present examples of global movements towards greater participation by children and youth.

Introduction

In this book we have argued throughout for a more relational understanding of the ways in which childhood, youth and the city are co-constructed. The preceding chapters have focused on the relationship between representations of space and age, structural issues that affect young people's urban lives in different parts of the world, and the spatial practices through which children and youth appropriate urban space. While the latter often occur in opposition to and in spite of urban policies and structures that marginalise young people, we examine in this chapter how the design of cities can be transformed through greater inclusion of children and youth in planning and policy processes. We examine a number of examples of child participation in urban planning and ask what such participation entails, how more active citizenship can be achieved and what problems need to be addressed in order to make participation thoroughly inclusive and effective. We examine the arguments for children's participation, describe a number of methods for achieving it, critically discuss the shortcomings of certain approaches and give a range of examples of participation in urban planning that connect with policy changes at different scales.

The context for change

Many of the changes in urban form, organisation, governance and practice that have occurred over the past few decades have had a negative effect on young people's rights to the city and to public space. The spaces available to children and youth in the city have become increasingly circumscribed due to developments such as rising levels of child poverty and increasing divisions of wealth, moral panics about young people *as* risk and fears about children *at* risk, or increases in traffic, the closure or privatisation of municipal infrastructure such as parks and playgrounds, and redevelopment strategies that erode rather than provide new spaces for children and youth.

At the same time, those working with and for young people have called for the creation of cities that are more inclusive and responsive to the needs of young people. As early as the mid-1970s, urban sociologists, child psychologists, planners and geographers began to pay attention to, and called for greater recognition of, the needs of children and youth in the city (Lynch 1977; Hart 1978; Ward 1978/1990; Matthews 1984). While such work did not, initially, achieve the desired shift in planning practice and research on the city, it heralded an important move towards greater participation and citizenship rights for children in urban governance, anchored in the 1989 United Nation's Convention on the Rights of the Child (UNCRC), which has to date been signed by most UN member states, but not the United States and Somalia. The UNCRC includes a number of articles that explicitly address the need for children's participation in governance. Thus, Article 3 declares that all actions concerning the child should take account of his or her best interests. Article 12 states that children have the right to express an opinion on all matters which concern them and that their views should be taken into account in any matter or procedure affecting the child, while Article 15 insists that children should have the right to meet with others, join and set up associations (children are defined as persons below the age of 18). Although far from being implemented thoroughly and effectively at all levels of governance in most states, the UNCRC has initiated a shift in thinking about the place of children and youth in society. At the level of urban planning, there has been an increase in initiatives to enhance children's participation and to work towards greater inclusiveness and responsiveness towards their needs. The 'child friendly cities' movement, initiated and coordinated by UNICEF, is one example of an internationally networked scheme that is applied differently by urban governments around the globe, but is inspired by the same goal of improving children's well-being in the city. At the national, regional and local level, there are now a multiplicity of other projects and policies to enhance children's inclusion, though they vary significantly in scope and extent from tokenistic, adult-led initiatives to thoroughly child-centred and child-led approaches.

Participation: planning cities *with* children and youth

Perhaps the most important reason why children and youth should participate in planning is that many decisions taken by planners affect them directly and their interests cannot be assumed by adults from the outside. Although they may need adult guidance on legal issues and areas of specialist knowledge, young people are experts of their own lifeworlds and from this unique perspective they will judge whether a planning project is likely to change their lives for the better or worse. While traditionally, planning evolved as a discipline of experts who were equipped with the necessary legal and practical knowledge, the profession increasingly recognises the importance of consulting with those most affected by its decisions (Levy 2010). Despite this, few planning manuals and textbooks include guidance on how to conduct participation and, more specifically, how to involve children. They continue to be excluded not only from general planning guides, but also from the practice of planning, due in part to perceptions of adults that children and youth are incompetent, unreliable, irresponsible and immature. It is assumed that children and youth have little if anything original to contribute, that they do not know 'what's best for them' and that their interests will inevitably clash with those of other community members. Taking young people's interests into account threatens those who see their presence as a disruption rather than as an enrichment of public life. This is one reason why, even in projects that consult young people, they are often 'designed out' (Eubanks-Owen 2002) by providing spaces on the margins of urban society and disregarding young people's claims towards visibility in central public places (Rogers 2006). In the words of Barry Percy-Smith (2002: 79):

> Young people appear to be caught in a systemic culture of non-participation in which democratic structures and processes remain exclusive, dominated by powerful interests such as development corporations, restricted by entrenched bureaucratic structures and paternalistic assumptions about young people's role in local decision-making.

Several authors have recently challenged these assumptions, however, and there is a growing body of knowledge on both the advantages and complexities of participation (Hart 1997; Nairn *et al.* 2006; Chawla 2002b, 2002c, 2002d, Driskell 2002; Freeman *et al.* 1999; Horelli 1998, 2007). Those who argue for increased participation by young people in planning and decision-making underline the following main benefits (see also Table 6.1):

- it enables a better response to the needs and interests of young people
- it increases young people's sense of belonging and responsibility
- it creates revitalised cities that respond to young people's needs and can benefit all

- young people develop valuable skills and learn active citizenship through participation.

Table 6.1 Benefits of young people's participation

Benefits for young people
• Participate in a new and exciting activity. • Look at and understand their local community and environment in new ways. • Learn about democracy and tolerance. • Develop a network of new friends, including community role models and resource people. • Develop new skills and knowledge. • Help create positive change in the local environment and other aspects of the community. • Develop a sense of environmental stewardship and civic responsibility. • Develop confidence in their abilities to accomplish the goals they set. • Strengthen their self-esteem, identity and sense of pride.
Benefits for other members of the community
• Interact with young people in positive, constructive ways, helping to overcome the misperceptions and mistrust that often exists between generations. • Understand how young people in their community view the world, their community and themselves. • Identify ways in which the quality of life for local young people can be improved. • Build a stronger sense of community and pride of place. • Appreciate the ideas and contributions of young people. • Invest time and energy in the future of the community.
Benefits for planners and policy makers
• More fully understand the needs and issues of the communities they serve. • Make better, more informed planning and development decisions. • Educate community members on the inherent complexities and trade-offs involved in policy and development decision making. • Implement at the local level the directions and spirit of the UN Convention on the Rights of the Child. • Involve young people in efforts to implement sustainable development, thereby helping to achieve the goals of Agenda 21 and the Habitat Agenda. • Create urban environments that are more child-friendly and humane.

Source: Driskell (2002: 35)

Adams and Ingham (1998) argue that young people are often more knowledgeable and concerned about environmental issues than adults. Environmental changes affect them most and they experience the environment differently from adults (also see O'Brien 2003). Their involvement can enrich the planning process, contributing a different set of values and priorities. It is beneficial for young people, since it prepares them to play an active role in shaping the environment as adults, but a 'generationally inclusive approach' (O'Brien 2003) can also improve urban life for all, not just for children and their parents (Cunningham *et al.* 2003). Children produce creative ideas for planning and their actions are frequently oriented towards more ecological and socially supportive communities. They direct policy planning to the long-term future and their involvement is crucial for long-term sustainability (Chawla 2002d).

For Freeman *et al.* (1999), the case for children's participation can be made on political, legal and social grounds. On a political level, the benefits that accrue from young people's involvement include improved services, a strengthening of participatory democracy, and the development of new opportunities, skills and insights for and by young people (Figure 6.1). Legally, greater participation is a requirement of the UN Convention on the Rights of the Child (UNCRC) (also see Skelton 2007), while socially, children and young people have fundamental rights to participate with everyone else as members of society.

Figure 6.1 Participatory exercise, Malawi. (Source: Lorraine van Blerk)

Learning task

Read the summary of the UN Convention on the Rights of the Child provided by UNICEF (http://www.unicef.org/crc/files/Rights_overview.pdf) and underline all aspects that require greater participation in decision making by children. Consider how participation rights are envisaged by the Convention and what obstacles there may be to their actual realisation. Further, ask which changes would be required to make cities in the Global South and Global North comply with the Convention. You may find it helpful to read Skelton (2007) in order to complete this task, and you may want to focus on two specific cities.

Moving beyond tokenism

Participation is seen by many authors as an important means of enhancing young people's citizenship rights (Matthews 2003; Cunningham *et al.* 2003, Chawla 2002d; Clark and Percy-Smith 2006). The geographer Hugh Matthews (2003) thus advocates *effective* participation by young people in political decision making and planning as a means of educating for citizenship, strengthening young people's status in relation to adults, and seeking greater recognition of their capabilities, interests and needs.

Yet, involving young people more consistently and genuinely in decision making means more than just listening to their voice. If their opinions and suggestions are to be treated on an equal basis to those of adults, then age-based relations of power also need to change. Greater participation by children and youth challenges existing presumptions and relations between adults and young people. It means thinking critically about whose interests are given priority and finding ways to negotiate between different interests. Yet, there are great variations in understanding and practice, as Chawla and Heft (2002: 202) rightly point out: 'When children enter the arena of planning for community development and environmental protection, they join a highly contested political realm where participation means different things to different people.' The authors caution that participation can be used as a 'system-maintaining' device, where information is disseminated in order to elicit cooperation, rather than as 'system-transforming', which involves the transfer of significant decision making to the people whose lives will be affected, even at risk of challenging existing structures of authority. While many projects offer good opportunities to interact and develop participatory skills, they fail to influence entrenched structures of decision making and are thus limited in reach and sustainability (also see Francis and Lorenzo 2002).

To make children's participation effective, a number of criteria should ideally be fulfilled. A comprehensive list of these was agreed upon by the Symposium on 'Children's Participation in Community Settings', organised by Childwatch International and the MOST programme of UNESCO at University of Oslo in June 2000 (Chawla and Heft 2002; Table 6.2).

In designing participatory projects, it is important to reflect on these criteria and to seek ways of creating the necessary conditions for achieving them. This does not mean that all of them will always be fulfilled, but they provide a useful yardstick against which to assess the effectiveness of participation.

A particularly important aspect to consider is the degree of participation that a project will offer those involved. Too often, young people are only involved in manipulative, decorative or tokenistic ways, which erode the very principles of engaged, meaningful and sustained involvement (Hart 1992, 1997). The children's geographer Roger Hart (1997: 41) instead suggests to work with the idea of a 'ladder of participation', where participation only really starts with 'assigned but informed' participation (step 4), and climbs from there to the following levels (see Figure 6.2).

- consulted and informed (step 5)
- adult-initiated, shared decisions with children (step 6)
- child-initiated and directed (step 7)
- child initiated, shared decisions with adults (step 8).

Not all of these steps need to be achieved and, depending on competence and interest, children may choose to participate at different levels. As Hart explains, '[t]he figure is rather meant for adult facilitators to establish the conditions that enable groups of children to work at whatever levels they choose' (1997: 41).

Freeman *et al.* (1999) likewise underline that, to be successful, projects need to be based on positive social attitudes towards children's participation, on positive adult–child relations and on a commitment to action. Projects need to have a realistic sense of what can be achieved. They should work with agreed ethical principles (Chawla 2002d) and confer rights on those participating (Freeman *et al.* 1999). Important questions to reflect upon in the design of participatory projects with children and youth include: how to involve young people in predominantly adult structures; how to balance children's and young people's issues against those of adults; how to empower children to develop their own alternative structures for participation; and how to work in partnership, seeing the benefits of and learning to engage in cooperation (ibid.). Ideally, participatory projects should create spaces where adults and young people can come together in dialogue, reflection and social learning (Percy-Smith 2006).

Table 6.2 Characteristics of effective projects for children's participation

Conditions of convergence
• Whenever possible, the project builds on existing community organisations and structures that support children's participation. • As much as possible, project activities make children's participation appear to be a natural part of the setting. • The project is based on children's own issues and interests.
Conditions of entry
• Participants are fairly selected. • Children and their families give informed consent. • Children can freely choose to participate or decline. • The project is accessible in scheduling and location.
Conditions of social support
• Children are respected as human beings with essential worth and dignity. • There is mutual respect among participants. • Children support and encourage each other.
Conditions for competence
• Children have real responsibility and influence. • Children understand and have a part in defining the goals of the activity. • Children play a role in decision making and accomplishing goals, with access to the information they need to make informed decisions. • Children are helped to construct and express their views. • There is a fair sharing of opportunities to contribute and be heard. • The project creates occasions for the graduated development of competence. • The project sets up processes to support children's engagement in issues they initiate themselves. • The project results in tangible outcomes.
Conditions for reflection
• There is transparency at all stages of decision making. • Children understand the reasons for outcomes. • There are opportunities for critical reflection. • There are opportunities for evaluation at both group and individual level. • Participants deliberately negotiate differences in power.

Source: Chawla and Heft (2002: 204)

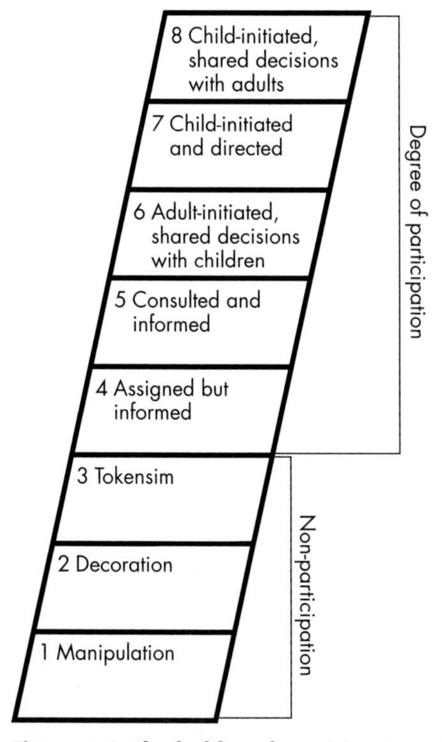

Figure 6.2 The ladder of participation. (Source: Hart (1992))

Participatory methods

In addition to applying the criteria and principles discussed above, participation can be enabled by applying a wide range of techniques that help children and youth with different abilities and interests to engage. Collaborative work should include 'a toolkit of a variety of methods, which allow the range of stakeholders to include even young people' (Horelli 1998: 236). Individual differences and special needs are as important to consider in the choice of methods as are social barriers to participation, such as gender stereotypes. Choosing a diversity of media and forms of engagement can help to overcome some of these barriers and to maximise the capacity for all children to take part (Hart 1997). A creative design of methods may also help participants to develop their competencies and to learn new skills. A wide range of examples are available from the literature. They are based on practice and experience, but most authors underline the significance of choosing methods appropriate to the project goal and to participants' abilities. It is also useful to consult artists, architects and youth work practitioners for new ideas and inspiration (see Sener 2006; Box 6.1).

Box 6.1 **The 'Obstacle Race' Project at Theatre Royal 2 in Plymouth, UK**

Figure 6.3 Mental map, Obstacle Race, Plymouth, UK. (Source: Kathrin Hörschelmann and anonymous project participant)

Figure 6.4 Self-directed photography, Obstacle Race, Plymouth, UK. (Source: Kathrin Hörschelmann and anonymous project participant)

In autumn 2003, the Education and Community team of Plymouth's Theatre Royal organised a series of workshops and events with primary and secondary school children (ca. 5 to 16 years of age), who lived in the district of Cattedown, where its main training centre had been opened only a couple of years earlier. The area of Cattedown lies in one of the most deprived wards in England and the award-winning building of the 'Production and Education Centre' of Theatre Royal (TR2) stands as a rather detached flagship development on the edge of the estate. The idea of the workshops was to bridge some of the divides between local residents and artists and to improve integration with the community. Participants in the project were recruited through a local youth centre and a nearby school. Girls and boys with a wide range of abilities took part in the project. Its main aim was to engage through a range of artistic forms with different aspects of living and growing up in Cattedown.

Workshops included dance, drama, drawing, music recording and photography (see Figures 6.3 and 6.4). They were led by staff trained in working with children and youth. All of the workshops were organised so that participants with different abilities could take part. While the staff suggested ideas and gave examples of how to implement them, most of the productions were led both in form and content by the young people. Starting with exploratory walks around the area, participants sought to express their experiences and impressions of living in Cattedown and in the city of Plymouth through their chosen artistic medium. In addition to representing well-known symbols of the city, they articulated a wide range of issues especially around fear of crime, experiences of vandalism, divided community, lack of play spaces and the sense of entrapment that many of the young people felt in the area. Few of them articulated positive experiences and had an answer to the question 'What do you like about where you live?' The project, however, allowed participants to articulate their concerns and to engage with adults in conversations about how to tackle some of the main issues raised. These conversations were sparked particularly by two final events, where the young people showed the results of their work to the wider community. In evaluating the project, participants emphasised that it had given them a valuable opportunity to engage with shared experiences of growing up in Cattedown, and that they had benefited from learning new skills, doing something different and building a stronger sense of belonging. Despite the buzz and excitement that the project generated, as a one-off event without clear links to policy and planning, its scope remained limited. However, it showed the value of working with children and youth through a range of artistic mediums, allowing a diversity of expressions and thus enabling young people with different interests and skills to actively engage.

Source: K. Hörschelmann, unpublished research

The following two books are highly commendable for offering hands-on, practical advice that is based on theoretical discussions of participation: Roger Hart's *Children's Participation* (1997) and David Driskell's *Creating Better Cities with Children and Youth* (2002). Hart recommends simulation and role play as a starting point. For those new to participatory work, this allows trialling it first and learning more about potential problems and pitfalls. Like numerous other authors, he underlines the importance of thorough research in order to ensure that children's actions grow out of their concerns and those of their communities. While arguing that traditional research has an important role to play, Hart advocates action research in particular since it is aimed at improving social conditions from the perspective of the researched, who become co-researchers.

Methods and techniques

Many of the methods that Hart describes have their origin in development projects, arising from the need to challenge top-down approaches and to respond better to community needs. In particular, Hart discusses the following methods (also see Box 6.2 and Chapter 2):

- drawings and collages: individual drawings, storyboards, collective drawing, collage making, drawings on slides
- mapping and modelling (see Figure 6.5)
- interviews and surveys – leading to trails and 'scored' walks
- media and communication – contribution to media/making media; dance/arts performances, festivals, parades, etc.

Hart recommends working with planning departments and other relevant institutions and organisations (libraries, city archives, schools) for resources. Thus, planning departments may have community maps that can be used as the basis for participatory activities and may be able to help out with technologies and materials.

One aspect of participatory work with young people that seeks to challenge existing conditions and to have a lasting, wider impact is raising public awareness through campaigns and engagements with the media. These may include public demonstrations, conferences, children's hearings, encounters with politicians and public officials, letters, public presentations, media reports and the production of their own media such as newspapers or videos. Hart warns, however, that children are too easily seduced into involvement in a movement which is not really 'their own' and that engagement with the wider public through media may distort their views, making them vulnerable to manipulation, decoration and tokenism.

Box 6.2 **Child-made maps**

For children under about eight years of age, it is best to allow them first to build a model of places they know well using wooden blocks, cardboard cutout houses and roads, and moss to serve as trees and bushes. Only after they have finished do I mention the word 'map', because sometimes this can intimidate them. In fact, all four-year-old children I have worked with are able to produce a recognizable map in this way. This can be done on the dirt or in sand, but at some point should be executed on paper so that the children can trace around their models, create symbols for them, and make a legend or key. The models are then removed, leaving behind their first map!

Making a map may initially intimidate children who have had little or no schooling with paper and pencil, but, once they begin, it is clear to them how much they know of their environment. The method can be particularly liberating for non-literate children, for it enables them to reveal to themselves and others that they have knowledge. Thus empowered through this informal method, they will be more likely to believe that they have other knowledge worthy of expression. The method can provide valuable insight for others into children's everyday environment because it is based on features that they consider important, and hence can lead to good discussion about aspects of their lives that might not so easily emerge in words.

Mapping personal worlds

For children who will be making maps for community research purposes, personal maps are useful as a first step. For children living in difficult circumstances, who need to investigate more and act upon the conditions of their life, such 'personal worlds' maps are fundamental. By mapping their use and evaluation of their daily environment, they can build a more ecological account of their world than if this information were collected by adults through an interview. In the case of street and working children, for example, such a map will include their social supports, activities, workplaces, and sleeping and eating places.

For children living or working on the streets, mapping is a useful way to express the history and current status of the spatial characteristics of their social world. Some of the problems in children's lives will relate directly to the spatial properties of this data, such as prohibitive transportation distances between home and workplace. Other spatial problems may be less obvious until one takes a close look at the maps with a child – for example, the conclusion that children are working in locations with the highest concentrations of carbon monoxide.

Source: Hart (1997: 165–170)

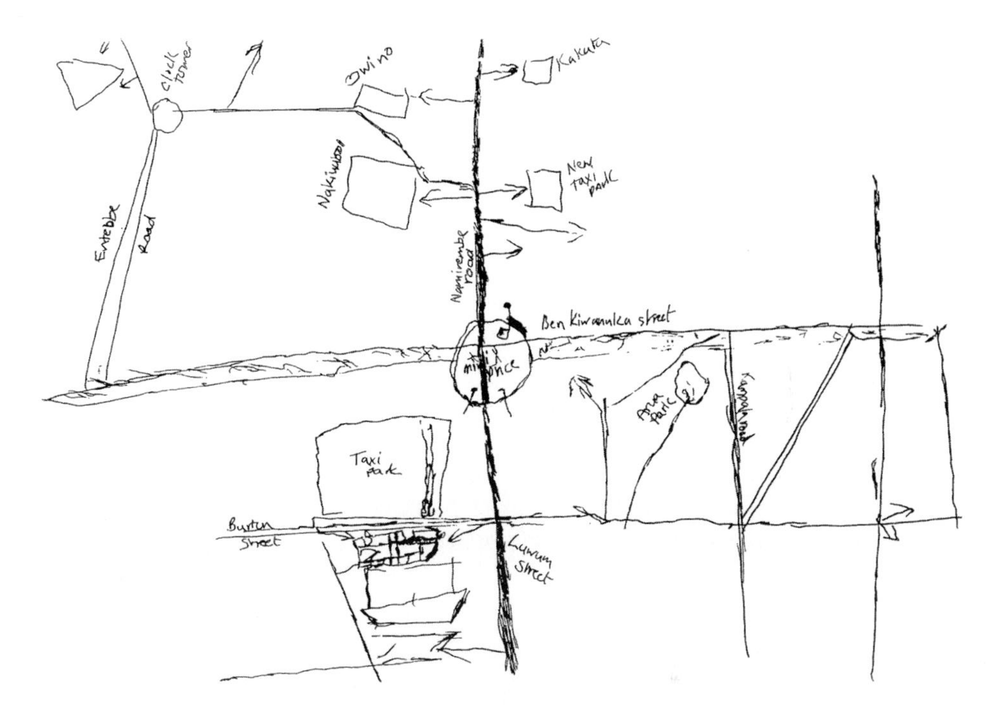

Figure 6.5 Mental map produced as part of participatory exercise, 17-year-old Ugandan boy. (Source: Lorraine van Blerk)

David Driskell's (2002) toolkit for children's participation is a more recent, equally user-friendly and reflective guide. It is based on the experiences of the 'Growing Up in Cities' project team, which conducted participatory action research with young people in seven cities around the globe (see Chapter 2; Chawla 2002a). Driskell (2002) describes the following activities that were conducted in this project:

- informal observations and 'hanging out'
- interviews
- drawings
- daily activity schedules
- documenting family and support networks
- role-play, drama and puppetry
- guided tours
- use of aerial photographs, maps and photographs of places
- photographs by young people
- behaviour mapping (to document use of a place)

- questionnaires and surveys
- focus groups and small group discussions
- workshops and community events.

These methods again provide a range of ways to promote active engagement, allowing participants with different skills and interests to become involved. They will, in all likeliness, result in a diverse set of findings and recommendations.

In order to agree on goals and priorities for planning, it is then important to prioritise and to develop an effective action plan. Driskell recommends the practice of 'visioning' for identifying common values, building consensus and developing a comprehensive statement of what is hoped to be achieved.

Box 6.3 **Guided tours, photographs by young people, Gulliver's mapping**

Guided tours

Tours that are guided by young people are consistently one of the most valuable methods for understanding their perspective on and use of the local environment. Viewing places first-hand often elicits new information and serves as a catalyst for more in-depth questions and discussion. To be of most use, guided tours should come after some of the other activities (such as interviews and drawings) when young people have developed a stronger relationship of trust with project team members, and team members have some initial information on the places they will be seeing.

Photographs by young people

Photographs taken by young people can be a valuable tool for gathering information on their environmental perceptions and attitudes, enhancing information collected through the interviews and other methods. Photographs can become the basis for discussions about the local area as well as providing visual data about it. They are also effective for initiating communication with the larger community through gallery displays. However, photography should not be proposed if it is considered culturally inappropriate ... If the equipment is accessible, short videos made by young people can be similarly effective.

Gulliver's mapping

This is a simple and fun activity that encourages residents of all ages to explore their memories, experiences and opinions about the local area. It can also be a highly visible activity that raises awareness and encourages dialogue about issues in the local environment.

A large map (1:250 to 1:5000 scale) of the local area is displayed and residents are invited to make any kind of comment they want on it. The activity takes place over the course of about ten days in a visible public location (such as a shopping centre or community centre). Ideally, the map is installed on the floor and people are invited to take off their shoes and walk or kneel on it to write their comments. There should also be wall space all around where additional information can be displayed or comments recorded, and enough room to accommodate dozens of people.

As comments begin to fill the map, close-up photographs are taken of the map space. These photographs focus attention on specific sites in the local area and their special meanings. These are Gulliver's Footprints – the individual and collective memories and meanings associated with places in the local urban habitat. The photographs, or Gulliver Cards, help to document the mapping results, and can serve as the basis for a photo gallery on the local environment as well as for site visits that can provide an opportunity to photograph the actual sites and to share more stories about each place.

Source: Driskell (2002: 127–171)

Another method that has been used in participatory research and can be applied equally effectively in planning with young people is story writing. Cunningham *et al.* (2003) give an example of a children's story-writing competition that was conducted as part of a broader exercise to develop a regional planning strategy for Blue Mountain City in Australia. The children were asked to write a story imagining that they were adults in 2025: what would their home and city be like? While Cunningham *et al.* found that this participatory technique delivered valuable insights that did not emerge from adult workshops, they also cautioned that different abilities needed to be accounted for. Great care had to be taken not to let the analysis be dominated by adult perceptions of the 'quality' of entries, but to focus on their content. This is an issue that arises with virtually all of the participatory methods mentioned thus far. The results of participants' work need to be interpreted in terms of the information they seek to convey, not their aesthetic quality or skill. Otherwise the contributions of some participants will be valued above those of others, thus reinforcing rather than challenging existing forms of exclusion.

Learning task

Identify a neighbourhood in a city that you know well and which you think would benefit from greater involvement of children and youth in planning and design. Develop a plan, either individually or as a group, for how you would conduct a participatory planning project in this area. How would you recruit participants? Who would you recruit? Which methods would you use? What would you need to consider if different age groups and young people with different backgrounds and abilities were involved? How would you analyse and present results with participants? What opportunities and which obstacles do you foresee for implementing changes recommended by your participants?

You may also be able to find existing participatory planning projects, either online or by contacting your local planning authority. In this case, you may be able to interview participants and facilitators about the methods they have used and their experiences with participatory planning.

New technologies and methods for participation

New computer technologies have, in recent years, inspired innovative participatory methods, including internet-assisted urban planning and participatory Geographical Information Systems (GIS) (Horelli and Kaaja 2002; Berglund and Nordin 2007). Internet-assisted urban planning can provide more space for young people to get involved in neighbourhood development, but it is currently used in a rather haphazard way (Horelli and Kaaja 2002). Many government websites, for instance, only include formal information and do not give opportunities for online interaction, thus limiting the potential for transforming representative democracy that could emerge from a more committed and effective use of the technology. Horelli and Kaaja (2002) argue that too often, discussions of young people's internet activities are limited by a technological determinism that either sees them as the avant-garde of technology use, ignoring issues of differential access and ability, or as threatened by the technology, reinforcing the perception of children and youth at risk (also see Bingham *et al.* 1999; Holloway and Valentine 2003). The authors present three examples of internet-assisted planning with young people in Finland, showing that their different visions and objectives led to great variations in the participatory potential of the technology. In the best case scenario, the interactive planning site of the Northern Finnish town of Joensuu enabled feedback between real and virtual events, where online interactions mobilized off-line engagements and visa versa. Young people, in this instance, were able to appropriate the technology for promoting their interests and for interacting and negotiating with other members of the community. Urban designs were made available online and a discussion-forum was created both off- and online for young people's comments. An internet-cafe has subsequently been opened at the central Joensuu Plaza, which is now used also as a Regional Resource Centre where young people network both online and off. Another example from Finland is the digital neighbourhood forum 'Home Street', developed by Helsinki University of Technology, which offers opportunities for interaction between local residents, authorities and businesses to communities in the Finnish capital (Box 6.4).

Box 6.4 **Home Street Project – net or trap**

Each school and library in Helsinki is on the net and a growing number of households are also connected to the web. The development of a digital neighbourhood forum, called 'Home Street', offers new opportunities for the management of cities in the information age.

The main goals of the Home Street Project are:

- to increase inhabitant participation in urban planning and design
- to innovate new managerial and planning instruments in communities and municipalities
- to strengthen the identity of a neighbourhood
- to underpin the existing social and economic capacity in the communities.

The Home Street Project has developed the internet as a participatory channel in urban processes. Interaction between the neighbourhoods and the municipality is determined from the perspective of the everyday lives of the inhabitants and not from that of the sectors and the organisations, as is usually the case. The pilot stage comprises the home pages of three different urban areas in Helsinki: Pihlajisto (3,000 inhabitants), Maunula (9,000 inhabitants) and Lauttasaari (25,000 inhabitants), which can be accessed via www.suomenkotiseutuliitto.fi/pihlajisto.

In addition to the general informative pages, the 'Home Street' comprises an interactive forum through which the inhabitants can make comments directly to the municipality, ask questions and propose improvements to their housing environment. The pages include different innovative tools for urban planning, such as a web based workshop for general planning, a web-based walkthrough for the planning of urban green areas and a digital neighbourhood photograph album.

Source: Aija Staffans (no date) http://www.greenstructureplanning.eu/MAPweb/Goteb/got-staff.htm (accessed 16/08/11)

Geographic Information Systems (GIS) can further be a useful tool for mediating between young people and planners and for fostering participatory dialogue, as Berglund and Nordin (2007) demonstrate. Working with several schools in Sweden, the authors developed a GIS application that was used by children and teachers to identify routes and places that children use in the city. The application entailed background maps, enhanced aerial photographs (that were removed after evaluation) and a questionnaire prompting participants to draw lines and symbols on the maps and to answer a set of questions. A version of this application was also used to examine traffic safety. Berglund and Nordin (2007: 479) conclude that: '"Children's maps in GIS" can provide a good start for planners who want to approach children with respect and make use of their technical and emotional competence […] in developing more inclusive urban planning.'

Participatory Action Research

Independent of the precise methods and techniques used, participatory planning projects benefit greatly from research, since otherwise they run the risk of replicating rather than challenging assumptions, excluding opinions and marginalising particular groups while lacking effective evaluation. Participatory Action Research (PAR) is particularly valuable in this respect, since it shares the same philosophy and can be integrated well into participatory planning projects. PAR aims to:

- develop a critical consciousness in researchers, participants and the broader community
- improve the lives of those involved in the research, through democratic processes
- transform societal structures and relationships that marginalise or disadvantage individuals or groups

- contribute directly to change in young people's conditions (Chawla and Malone 2003: 128).

Research thus becomes a vehicle for social change (Pain 2004; Kindon *et al* 2007), a space for questioning exclusionary practices and social inequities, and a means for participants to look critically at their social and/or environmental contexts, enabling them to develop ways of addressing the problems they identify (Cahill 2004; also see Beazley *et al.* 2009; Eversole 2009; Kesby 2005). PAR starts and ends with the perspectives of participants, but instead of stopping at an investigation of their existing views and opinions, it enables critical engagement and becomes a process for challenging perspectives and moving towards actions that respond to needs identified in the *process* of research. An excellent example of how such an approach can be applied to challenge marginalisation in the process of urban renewal is Caitlin Cahill's (2004) work with young women growing up in the Lower East Side neighbourhood of New York City. Cahill's feminist approach is based on the ideas of Paolo Freire. It starts with the concerns of young women and their own critical investigation of their social contexts, moving from emotional motivation through the politicization of personal experience to speaking back with research. Cahill works critically with and through her own position as a white, young woman rather than ignoring how her positionality influences the research. She describes the painful process of identifying and naming exclusionary practices, the emotional investments of her research participants in seeking to challenge these exclusions and the energy with which they resolve to 'speak back', using a sticker campaign, website and research report.

The use of PAR to challenge the marginalisation of particular groups of young people and to change their social conditions is also well demonstrated in a Participatory Action Research project with street children in Turkey (Ataöv and Haider 2006). In Turkey, as in many other countries around the globe, street children are treated as 'undesirables' and a 'social threat'. They are rarely acknowledged as actors in public space. The PAR project with street children aimed not only to challenge these negative assumptions, but also to empower children and create a learning environment 'in which children could systematize their real-life situations, formulate the best solutions to address their problems, and act upon decisions' (ibid.: 132). The researchers used art, cognitive mapping, graphics of street children's support networks, dramas, structured and semi-structured interviews, maps of migration, peer-interviews, and in-depth discussions of life-stories in order to explore with street children in different Turkish cities how they had come to live and/or work on the street and how it affected their social positioning and well-being.

Another large-scale research project that used participatory methods in order to produce a better understanding of the needs of urban children and youth from their perspective is the 'Growing Up in Cities' project (GUIC). GUIC was originally developed by Kevin Lynch in the 1970s and recently resurrected by Louise Chawla in the late 1990s (see Chawla 2002a). It is an interdisciplinary, cross-national project that both seeks to understand urban environments from children's perspectives and develop children-adult partnerships for improving local communities by including children's voices in environmental planning and decision making. GUIC had two important and overlapping research aims:

1 To understand how young people use, value and seek to improve their urban communities.
2 To understand how to move governments and leaders in civil society to enable young people to participate in constructive urban change.

GUIC took place in eight diverse countries and aimed to develop research-based, participatory urban planning and design with children and youth. By highlighting the case studies from Norway and South Africa, it is possible to see how this ideal of participatory research was clouded by popular adultist representations of childhood in the respective countries. In Norway municipalities are required to appoint a children's representative to defend children's interests in relation to planning cases connected with land use. This meant GUIC had a good starting point for bringing children's interests into urban planning, the two contrasting case study areas of Mollenberg and Finalen, and how planners dealt with children's research ideas, demonstrate however that they were influenced by popular ideas about a good childhood which in Norway links children with nature – creating outdoor spaces for free play. In Finalen where a local park was threatened due to expansion of the hospital, children's requests were noted and the park was saved, whereas in Mollenberg, children felt they had plenty of park space and made suggestions to reduce traffic and increase safety in the streets. A green space that the young people considered valuable as a wild play space did not fit with the planners' ideals of where children should play and was subsequently built on without further consultation. The popular conceptualisation of childhood as a time for innocence and learning (akin to Jenks' (2005) Apollonian vision of the childhood ideal) influenced the way in which children's voices were utilised and represented.

Turning to the South African case study (Griesel *et al.* 2002; Kruger and Chawla 2002), there are some parallels with the Norway example in that adult understandings of childhood had an impact on the young people's planning research. The case study was Caanansland squatter community in Johannesburg – an area stigmatised by outsiders and considered dirty and disease ridden. There was only one water tap for thousands of residents and no sanitation or electricity. Through their involvement in the project the children's self-esteem and self-efficacy improved and they were able to draw up a list of improvements based on their perceptions of the community. These included toilets; piped water; better garbage disposal; stronger shacks; fencing to keep out criminals; a space to study; and electricity. GUIC secured good support for the project and the mayor held a workshop to discuss the children's proposals. Although this seemed very positive, the squatter camp was relocated without notice; outsiders to the community had complained that it was a blemish on the landscape. The children's views were not sought or implemented.

Figure 6.6 Slum area in Cape Town, South Africa. (Source: Lorraine van Blerk)

Addressing power relations in participatory action research

Young people's participation in research has been advocated as useful for ensuring their voices are heard. Similarly to the GUIC project, where young people were included in the design, implementation and dissemination of the research, researchers investigating other areas of young people's lives in the city have adopted a participatory process for overcoming power relations and for ensuring methods are appropriate for the particular age and competencies of the children involved. Some projects have empowered young people through training them to carry out the research themselves (Porter and Abane 2008; Cahill 2004), while others have included children's ideas and views on various aspects of the research from the topics to be researched, the methods and techniques used and where and how the process should take place. The usefulness of participatory research should, however, be examined critically and not necessarily considered without challenges to the researcher. Schäfer and Yarwood (2008) highlight that even where young people are empowered to carry out research themselves, this does not necessarily overcome issues of power and powerlessness as unequal power relations inherent within groups of young people can significantly influence the research process.

Other approaches to researching children and youth have involved discussion around the actual methods and techniques employed. There has been much discussed regarding the adaptation of traditional research methods and the development of more innovative child-centred methods (Bingley and Milligan 2007; Langevang 2007; Valentine *et al.* 2001; Young and Barrett 2001), both qualitative and quantitative, for increasing participation and reducing power relations in the research process so that children are given a more equal voice. As Valentine (1999b) points out, because childhood and youth are defined in contrast to adults, researchers can place them in positions of powerlessness through the strategies and approaches they use. This has led to researchers arguing for 'child-friendly' methods to be used in research with young people to overcome these situations and techniques such as drawings, maps, photographs and diaries have become popular among those researching the experiences of young people in the city (see Dodman 2003 and Langevang 2007 for examples).

However, Punch (2002) argues that researchers must be careful not to overemphasise differences between adults and children, raising the question should research with children use different methods? Punch (2002) argues that those who see children as the same as adults often engage in the same methods they would use with adults, to the detriment of the research process. They ignore the imbalance of power relations between children and adults, and in fact reinforce these. However, researchers who see children as inherently different to adults and develop specifically 'child-friendly' methods may end up patronising young people and undermining their competencies. Consequently, researchers working in this area agree that using a combination of methods is useful – both new innovative methods and adapted traditional methods derived from a research philosophy of respect for children. This includes discursive methods such as interviews and focus groups as well as written and visual methods, which depend more on young people taking active control of the task at hand.

In addition to all the qualitative methods advocated for researching children, youth and the city, Boyden and Ennew (1997) argue that we should not dismiss the value of quantitative research, although questionnaires have been heavily criticised for their authoritarian approach. They argue that questionnaires can add value to a research project providing valuable quantitative data as long as they are administered correctly. This usually means running the questionnaire towards the end of a research project when young people are familiar with the topic and the researchers. Porter and Abane (2008) used children to administer the questionnaires to reduce the authoritarian stance of the method.

Finally, it is helpful to highlight that there may be particular ethical and practical issues related to undertaking participatory research with children and youth in the city, and this is particularly important when researching with minors under the age of 18 (Valentine 1999b). There are numerous ethical codes of conduct and practice produced by different academic disciplines and bodies but the key issues to bear in mind in your own research should begin with a principle of 'doing no harm'. This means that it is important to think through the benefits of undertaking research that might be considered intrusive or dealing with difficult subjects and in such circumstances, if deemed appropriate, how to minimise any potential distress. Fully informed consent is essential, whereby the project is explained to young people in a language they can understand and they are then given the option to participate (for more detailed consultation of ethical issues in research see Valentine 1999b).

Obstacles to and limits of participation

Despite many successful examples of participatory planning with children and youth, there is a number of problems and limitations that practitioners and researchers of participation have identified. From a planning perspective, social inequalities that arise from uneven distributions of wealth are difficult to tackle through planning means alone. They require a sustained political effort to redistribute resources and are not sufficiently addressed by measures to redesign urban space with and for children and youth. Initiatives such as the UNESCO programme for 'Child Friendly Cities' therefore combine participation in planning with demands to tackle social inequalities and improve the public infrastructure that children need for their health and well-being (Bartlett *et al.* 1999). Nonetheless, young people's active involvement in planning decisions can help to raise attention for issues of social marginalisation and to exercise pressure on decision makers. Designing urban spaces with the aim of making them more inclusive is also an important part of addressing issues such as unequal access to outdoor green areas and play facilities.

Effective and sustained participation is limited by a number of other negative factors. Thus, although the benefits of participation are beginning to be recognised in the planning literature (Levy 2010), few architects, planners and designers have been trained in participatory techniques, consider the specific needs of children and have knowledge about their use of public space (Freeman *et al.* 2003; also see Valentine 2004). Children's participation in planning and in policy decisions is hindered particularly by the persistence of negative adult attitudes. Young people continue to be seen as less competent, as irresponsible or as bored and disinterested, even though research has attested time and again that they wish to be consulted on issues that concern them (Matthews 2003; Matthews *et al.* 1999; Adams and Ingham 1998; Douglas 2006; Freeman *et al.* 2003). Adults may also feel that children should be shielded from the burden of responsibility (Matthews 2003; Morrow 1999) and assume that they know best, even when it comes to the interests of children and youth. When consultation takes place, it tends to be conducted for adult purposes and with techniques that are perceived to work well with adults, thus making little attempt to adapt to the needs and interests of young people (Cunningham *et al.* 2003). It can be difficult for adults to let go of the power that places them in positions of advantage and they need training in order to learn how to engage in open and genuine dialogue with children and youth (Matthews 2003; also see Kelley 2006). Tracey Skelton, in her review of the 2003 UNICEF Report on 'The State of the World's Children' likewise critiques the reluctance, even by an organisation that promotes children's participation, to transform rather than confirm existing models of democratic engagement. She asks:

> Is something of the vitality and creativity of children and young people lost when they participate in adult structures? If pre-existing models have marginalized children then unless there is fundamental change within the institutional structures children's participation will appear as tokenism, no

> matter how often this accusation is denied. Just as men were, and are, reluctant to give up their established forms of political (and other types) of power to allow women to play a meaningful role, so adults will resist the loss of authority and power that a child-centred, young person-friendly model of democracy will require.
>
> (Skelton 2007: 178)

Consultation with children is also too often seen as 'special'. It tends to be driven by perceptions of conflicting needs between adults and children (Malone 2002) and is usually conducted in isolation from the broader community (Cunningham *et al.* 2003). Opportunities for dialogue are thus restricted and, without sufficient feedback and accountability towards young participants, the dialogue becomes tokenistic and potentially exploitative (Kelley 2006; Percy-Smith 2006). Barry Percy-Smith (2006) therefore argues that more attention needs to be paid to how young people and adults can participate together. He suggests that intergenerational communication is best achieved through collaborative social learning, emphasising again that both adults and young people enter the participatory process as *learners*. Hill *et al.* (2004: 80) likewise emphasise the importance of genuine dialogue, for '[o]nly if genuine dialogue occurs between children and the adults in power will policies directed at social inclusion respond to children's felt needs, rather than to needs attributed to them'.

Even where children and youth are thus consulted, their ability to contribute to broader change is often limited. Elsley (2004) describes the case of a Scottish regeneration project in an Edinburgh housing estate that ranks high on list of areas of deprivation in Scotland. The young people she interviewed complained that adults outside of their personal networks did not to listen to their concerns. They found that their needs for better leisure provisions on the estate were regarded as a low priority by adult planners and not taken seriously enough. Rather than being involved in the full process of planning and decision making, their participation was compartmentalised and focused exclusively on youth specific issues.

Compartmentalisation and young people's segregation from adult society (Ennew 1994) is a major limitation on the wider reach and effect of their participation. Too often, children and youth are consulted only about issues that appear to be specific to them rather than being involved in the whole planning process (Cunningham *et al.* 2003). Participation that is limited to youth-specific issues can thus, paradoxically, become a means to further control their activities and to reinforce their exclusion from public space (Prout 2000b). Hart (2002) shows how this has happened through the design and privatisation of play spaces in New York, which has further separated children from the daily life of their communities and fails to address the complexity of their needs. Like Colin Ward (1978/1990; see Hart 1997), he argues that the city needs to be transformed as a whole rather than creating only niche spaces for young people. Instead of more segregated playgrounds, greater efforts need to be undertaken to make neighbourhoods safe and welcoming for children and to offer them opportunities to explore and play freely. This argument is shared by Hill *et al.* (2004: 84), who emphasise the need for a shift in adult thinking about the spaces that children inhabit:

> It is important that adult thinking about children's lives, needs and education embraces not only the spaces to be found in formal provision by adults, but also those territories and pathways claimed by children for their own purposes in myriad locations within the areas they inhabit and visit.

Much of urban planning nonetheless continues to focus on ways of designing young people out of public spaces. Youth specific spaces such as skate parks are thus created at least in part as a way of alleviating tensions between adults and young people over the use of public space (Malone 2002; Prout 2000b). Rogers (2006) gives an example of this based on his research in the city of Newcastle, UK. Despite being consulted, young people were here designed out by urban renaissance policies that are framed primarily around economic growth agendas and not social renewal strategies. Rogers argues that 'renaissance' potentially masks an agenda of exclusion through policies and provisions that seek to provide youth-specific redevelopments. Young people are demonised in public British discourse as 'gangs' and 'yobs'. They are seen as a problem to be solved, not as a group with valid claims on public space. In Newcastle, young people who met at one of its most popular sites for youth interaction, Eldon Square, were consulted by means of a questionnaire about their redevelopment interests and priorities. Although a large proportion of responses (32 per cent) were so fragmented that they were discarded in the analysis and only 17 per cent of young people requested a skate park, this was the suggestion taken forward by city planners. While young people were subsequently consulted about the design of this space, it was located on the northern edge of the city centre rather than more centrally, as requested in the survey. The outcome of this exercise has thus been a further segregation of young people from public space, neglecting calls for a centrally located youth centre, and a privileging of the interests of only a small group of young people, partly because the use of central public space by skaters was seen as particularly problematic. No dedicated youth space has been created and there has been little recognition that different youth cultures use urban space in distinct ways:

> In this situation, young people have been treated as passive subjects unless directly addressed by intermediaries. As a result, they have been subject to a wide range of efforts made by managers on their behalf, but with an

> undercurrent of an intention to relocate them to a more appropriate facility. This emphasis on relocation implicit in managerial policy in effect undermines the intentions of youth participation in the generation of policies for urban redevelopment.
>
> (Rogers 2006: 118)

The contradiction between consulting and excluding young people that is evident in this example is a reflection of broader tensions between public policies and practice that recognise children as persons in their own right and seek to promote their social inclusion and citizen participation on the one hand, and increasing surveillance and control of young people on the other (Prout 2000b):

> Policy responses have thus given out paradoxical and ambiguous messages with regard to young people's participation. On the one hand, young people are lauded as 'active citizens' who can make valued contributions, on the other hand, the state is extending an authoritarian hand of regulation and control in the management of public space and young people's use of it.
>
> (Percy-Smith 2006: 159)

Rather than shying away from these problematic issues, more critical reflection on the effectiveness of participation is needed. Matthews (2001, 2003) and Matthews *et al.* (1999) for instance identify a lack of transparency about political processes as a key obstacle to achieving effective change through participation. Young people may be offered power but find themselves with little as decisions are taken behind the scenes. This can lead to disillusionment, which discourages further participation:

> Unless young people are confident that their opinions will be treated with respect and seriousness, they will quickly become discouraged and dismiss the participation process as ineffective, with all the implications this has for their confidence in democratic processes as they grow into adulthood. We suggest that poor participatory mechanisms are very effective in training young people to become non-participants.
>
> (Matthews *et al.* 1999: 140)

Participatory projects are also often organised as one-off events that are thus unlikely to inspire sustained, long-term change and to reach beyond youth-specific issues (Sinclair 2004; Percy-Smith 2006). As Matthews (2003: 207) argues: 'Lack of sustainability together with the short-term needs of local officials to tick the correct social measure as part of their performance review are fundamental weaknesses of many community rehabilitation schemes.' Authors such as Matthews (2001, 2003), Percy-Smith (2006), Freeman *et al.* (2003) and Kelley (2006) consequently call for a move from short-term policy developments to longer-term shared engagements that enable adults and children to think and act collaboratively. This means creating spaces for the development of shared understandings, greater commitment to involving children and youth as equal partners right from the start of regeneration projects and other initiatives. It means consultation and involvement at all stages, transparency about procedures and outcomes, clear feedback mechanisms and continued dialogue. Adult support is crucial for this since, as 'outsiders', young people need their advocacy and knowledge of institutions and procedures (Matthews 2003; Adams and Ingham 1998; Freeman *et al.* 1999; Chawla 2002a). Yet, at the same time adult professionals need to reflect more on their own values, systems and priorities and learn how to give up privileges that hinder dialogue and participation (Percy-Smith 2006).

A further significant issue to reflect on in participation projects is the diversity of young people, whether in terms of their social and cultural background, their competencies or their interests and needs (Hart 1997; Driskell 2002; Nairn *et al.* 2006; Harris 2006; Cunningham *et al.* 2003). Young people are not one, well-defined, homogenous group (Adams and Ingham 1998), yet too little effort is often made to think carefully about ways in which different groups may become involved and to challenge universalising and normative assumptions about childhood and youth (Sinclair 2004; Skelton 2007). Adult facilitators may have preferences for particular groups (Nairn *et al.* 2006; Hill *et al.*2004), while not all young people choose to participate (Matthews 2003) and not enough effort is made to work with those who are classed as 'hard too reach' (Cunningham *et al.* 2003). Participation initiatives may thus 'reinforce existing patterns of social exclusion and disadvantage' (Lowndes *et al.* 2001: 453). Nairn *et al.* (2006) also demonstrate, however, how the tendency to involve either the most motivated or the most 'troublesome' youths leads to the exclusion of large numbers of young people 'in the middle'. Examining local authority practices in New Zealand, they note that local government staff are most interested in working with 'achievers', who are recruited via schools, and with possible 'troublemakers', identified by youth workers. It is the former who then tend to represent all young people in youth councils. Representatives of minority groups are invited to participate in a tokenistic way, meaning that only 'exceptional' Maori, Pacific Island and/or gay young people come to represent these diverse groups. The dominant middle-class environment works to exclude non-white and non-middle-class youth councillors and organisations such as school often still play a major role in mediating diversity.

The diversity of children and youth's competencies and skills also needs careful consideration if participatory projects are to be inclusive and responsive to different abilities. Using a wide spectrum of methods and techniques is one way to achieve this, as is the creation of an open, supportive and mutually respectful atmosphere. Children with disabilities, however, may have further requirements that need to be addressed in order to ensure that they can participate on an equal basis and adult facilitators need to be trained to respond to these requirements (see Cosco 2002; Box 6.5).

Box 6.5 **Extending the participatory process to children with special needs**

In planning sessions to include children with disabilities, consider the following:

- Send out a message of universal acceptance. Make clear that *all* abilities are welcome.
- Ask families or community groups to let you know in advance if children with special needs will be attending your programme.
- Consult with experts in different disabilities to prepare adequate materials.
- Sign language interpreters and cards with graphic instructions might be helpful.
- Use simple language, or other artefacts to assist communication, and processes such as structured role playing and games.
- During meetings, sit at the same level as the children.
- Open meetings by welcoming the participants, introducing them by name, and present the steps that you are going to follow. Make sure that the children with disabilities have properly understood the process.
- Allow interaction of siblings as helpers/interpreters. Ask them what is the best way to communicate with the child if you are insecure.
- Provide a calm atmosphere, without background noise.
- Concentrate your questions on two or three main issues.
- Remember that instructions and questions must be simple and direct.
- Allow enough time to respond to questions.
- Do not hesitate to ask the child to repeat their answer to aid clarity of expression.
- If a child has a hearing impairment, be sure there is no background noise, that the child is able to see the faces of other participants, and pay attention to gestures and signs that the child may make.
- Children with physical disabilities may feel tired quickly. Be ready to give a break or to reschedule the session.
- For a child with visual impairments, prepare the room without obstacles and conduct the meeting in a well-illuminated area. Avoid glare.
- Always allow enough time to establish rapport.

Source: adapted from Nilda Cosco (2002: 58f)

Making the city a good place for children to live

Cities are frequently seen as problematic places to bring up children. Yet, a well-planned city can have many advantages for young people and their families (Chawla 2002c). There is a better availability of services and they are in closer proximity. The concentration of people brings with it greater opportunities for encountering difference and finding a wide range of friends, and the whole buzz and noise of the city may be attractive for young people. For urban managers, on the other hand, the higher concentration of services, infrastructure and people allows a better focusing and use of resources. It should also, ideally, make it easier to implement environmentally friendly policies.

Cities that are friendly to children and youth can, of course, not be defined in abstract and generally valid terms. This is why we have emphasised the significance of participatory projects that respond to the needs of diverse groups of young people. However, participatory research with young people in cities around the globe has delivered some important insights that appear to be more universally applicable. They are based on the views expressed by young people themselves and start, fundamentally, with the need to challenge the current construction of the city as an adult space that outlaws and ghettoises children (Ennew 1994). Young people's different needs and interests have to be recognised and their rights to have a say on issues that concern them acknowledged (Freeman *et al.* 1999; Valentine 2004).

Learning task

In order to consider what aspects would make a city a good place for children, David Driskell (2002) has developed two simple and interrelated exercises. He suggests first of all to answer the following set of questions:

> *Is My City a Good Place for Young People?*
>
> What do the young people in my city feel about the place where they live? Do they feel valued as members of the community? Are there places for them to meet with their friends? Do they participate in community activities? Do they feel safe? *Is my city a good place in which to grow up?*
>
> (Drsikell 2002: 26)

Driskell then recommends asking whether '*you* think your city is a good place in which to grow up? What do you think are the most positive features of your city, from a young person's perspective? What do you think are the least positive features?' (ibid.: 27). The answers to these questions should be compared with those of young people and followed by a critical evaluation of the differences between one's own answers and those of youth participants:

> Were there things identified by young people that did not appear on your list? Did they emphasize issues that did not seem so important to you? Did you include issues that young people did not see themselves? (ibid.: 27)

Based on the Growing Up in Cities project, Driskell (2002) and Chawla (2002c) suggest a number of indicators of environmental quality from children's perspectives (Table 6.3). High up on the list of positive indicators are social integration, cohesive community identity and traditions of self-help, which can help to ameliorate social disadvantages in places otherwise classified as 'poor' (Bannerjee and Driskell 2002). Social exclusion, stigma, violence and crime are the main negative indicators, showing how social instability affects young people's well-being. While the significance of indicators varies between places, they offer a useful guide for evaluating the child-friendliness of a city.

There have been few such attempts at defining standards of children's rights in the built environment, and while many commercialised spaces now include service facilities that are adapted to children (child-sized toilets and baby change areas, for instance), 'there is no requirement to provide child-sized toilets, for doors to have handles at a suitable height and which are not too heavy to be opened by children, for children to be housed at ground-floor level in multi-storey housing' (Freeman *et al.* 1999: 121). Limited resources and the dominance of adult interests mean that urban governments rarely prioritise these issues. Freeman *et al.* (1999) suggest that this will only change if pressure is put on politicians and decision makers to orient resource allocation and investments more strongly towards the future. While emphasising that there is no one 'right' environment, they recommend that the following environmental rights would contribute to more child-friendly environments and greater equity between the needs of adults and children:

- acknowledge legitimacy of children's environmental rights
- include right to participate in determining environmental use and design access to whole city or countryside environment
- being part of the community
- having access to natural environment
- being able to use and reasonably access public services
- physical environments and public services that recognise children's specific needs

Table 6.3 Indicators of environmental quality from children's perspective

Positive indicators	*Negative indicators*
Social integration: Children feel welcome and valued in their community.	**Social exclusion:** Children feel unwelcome and harassed in their community.
Cohesive community identity: The community has clear geographic boundaries and a positive identity that is expressed through activities like art and festivals.	**Stigma:** Residents feel stigmatised for living in a place associated with poverty and discrimination.
Tradition of self-help: Residents are building their community through mutual aid organisations and progressive local improvements.	**Violence and crime:** Due to community violence and crime, children are afraid to move outdoors.
Safety and free movement: Children feel that they can count on adult protection and range safely within their local area.	**Heavy traffic:** The streets are taken over by dangerous traffic.
Peer gathering places: There are safe and accessible places where friends can meet.	**Lack of gathering places:** Children lack places where they can safely meet and play with friends.
Varied activity settings: Children can shop, explore, play sports and follow up other personal interests in the environment.	**Lack of varied activity settings:** The environment is barren and isolating, with a lack of interesting places to visit and things to do.
Safe green spaces: Safe, clean green spaces with trees, whether formal or wild, extensive or small, are highly valued when available.	**Boredom:** Children express high levels of boredom and alienation.
Provision for basic needs: Basic services are provided such as food, electricity, medical care and sanitation.	**Trash and litter:** Children view trash and litter in their environment as signs of adult neglect for where they live.
Security of tenure: Family members have legal rights over the properties they inhabit either through ownership or secure rental agreements.	**Lack of provision for basic needs:** When basic services such as clean water and sanitation are lacking, children feel these deprivations keenly.
	Insecure tenure: Children, like their parents, suffer anxiety from fear of eviction, which discourages investment in better living conditions.
	Political powerlessness: Children and their families feel powerless to improve conditions.

Source: Chawla (2002c: 228–229)

- designing public spaces which take into account children's needs
- being able to move about freely
- being able to travel safely
- having areas designed specifically to meet children's needs.

These rights focus on both physical and social aspects, since children's environments include urban form as well as the social environment.

Freeman *et al.* (1999) echo Hart (1997) and Ward's (1978/90) argument that investment into children's urban environment should not be restricted to dedicated zones or play areas, but that the city as a whole ought to be transformed. One example of such a shift in attitudes and priorities are Dutch 'wonnerfs', or 'home zones', where traffic is calmed by cul-de-sac roads and the need for play areas is placed before the interests of car-drivers (see Figure 6.7). A spinal footpath network and informal play spaces increase both children's mobility and their safety within neighbourhoods while making allowances for creative, unsupervised play. Freeman *et al.* (1999) argue that children need to be able both to experience and to manipulate space. They should have opportunities to put their creative stamp on places rather than finding their activities constrained by manicured, formal designs that are often a response to excessive safety fears rather than children's developmental needs (Hart 2002). Developers of play spaces and outdoor green areas increasingly incorporate these ideas into their designs by making allowances for 'non-directed' activities and intergenerational encounters.

Figure 6.7 Play area in an inner-city neighbourhood of Groningen, The Netherlands. (Source: Kathrin Hörschelmann)

From the global to the local: child-friendly cities

Efforts to improve the situation of children and youth in cities have taken place across different scales over the last two decades. Whether at local or international level, most have been inspired by and respond to the UN Convention on the Rights of Children (UNCRC) with its call for greater participation rights of children. At the international scale, the Child Friendly Cities movement has been the most significant attempt of stimulating and coordinating national and local projects to improve urban children's well-being and social inclusion. The movement aims to translate the commitments made by national governments under the UNCRC into action at the city level. As Riggio (2002: 45) explains:

> The concept of 'child friendly cities' has been developed to ensure that city governments consistently make decisions in the best interests of children, and that cities are places where children's rights to a healthy, caring, protective, educative, stimulating, non-discriminating, inclusive, culturally rich environment are addressed.

Efforts to involve municipal authorities in the implementation of children's rights started with the Mayors Defender of Children initiative that was launched in Dakhar, Senegal, in 1992. The notion of child-friendly cities emerged from the 1996 UN Conference on Human Settlement, Habitat II, in Istanbul (Riggio 2002; Bartlett *et al.* 1999). Since then, there have been several international conventions and in September 2000, an International Secretariat for Child Friendly Cities was set up. The task of the secretariat is to provide a common framework for child-friendly cities and to support the growing network of city authorities and other groups involved in child-friendly initiatives by gathering information, carrying out research and promoting good practice.

Child-friendly city initiatives are guided by four key principles: non-discrimination, the best interests of the child, the right to life and maximum development, and respecting children's views. They promote active participation by children and deal with young people's needs holistically. In order to achieve the implementation of the UNCRC at the municipal level, child-friendly cities engage in institutional, legal and budgetary reforms.

Without being rigid or prescriptive, child-friendly city initiatives aim to give visibility to the child in the city-development agenda and to grant children opportunities to participate in decision-making process. They are concerned with the whole range of human rights for all of a city's children (Riggio 2002: 48):

> Fundamentally, a child friendly city aims to guarantee the right of all young citizens to:

- Influence decisions about their city;
- Express their opinions on the city they want;
- Participate in family, community and social life;
- Gain access to basic services such as health care, education and shelter;
- Drink safe water and have access to proper sanitation;
- Be protected from exploitation, violence and abuse;
- Walk safely in the streets, on their own;
- Meet friends and play;
- Have green spaces for plants and animals;
- Live in an unpolluted and sustainable environment
- Participate in cultural and social events;
- Be supported, loved and cared for; and
- Be equal citizens with access to every service, regardless of ethnic origin, religion, income, gender or disability.

A particularly interesting and inspiring example of what has been achieved in some countries under the child-friendly cities framework are the children's participatory budgets developed in Brazil. Guerra (2002) examines the case of Barra Mansa, a city of 170,000 inhabitants located between Rio de Janeiro and Sao Paolo. The city has implemented a structure for children's participation that evolves from the neighbourhood level right through to the level of municipal planning. Children participate in neighbourhood assemblies and send delegates to district assemblies which then, in turn, elect 18 boys and 18 girls between 9 and 15 years as child councillors. The child councillors determine how a proportion of the city's budget is spent on children's priorities that have arisen from debates in the neighbourhood and district assemblies. The councillors also undertake visits to different neighbourhoods and thus develop a better understanding of variations in living standards, different children's needs and the variety of interests that they need to address.

The participatory budget has allowed the child councillors to fund a wide range of projects such as repairs to schools and school equipment, better security, improved playgrounds in low-income areas, repairs of sewers and drains and tree planting projects.

In a European context, positive examples of schemes for the promotion of children's rights that have led to the implementation of child-friendly principles at city level include the Children's Ombudsman initiative, which was started in Norway in 1981, the promotion of youth parliaments and youth councils, the 1992 'European Charter on the Participation of Young People in Municipal and Regional Life' and the European Youth Directorate (Matthews *et al.* 1999; also see Wilhjelm 2002). Matthews *et al.* (1999) cite the children and youth town councils in France as exemplary places for the expression of young people's values and concerns and as places where young people can acquire the skills of citizenship. A similar scheme of youth councils in towns and cities has been implemented in Italy, a country that is lauded by several authors for being at the forefront of implementing the child-friendly cities agenda (Baraldini 2003; Francis and Lorenzo 2002; Corsi 2002; see Box 6.6).

While the US has not signed the UN Convention for the Rights of the Child, cities such as Denver subscribe to the principles of the child-friendly cities initiative and have implemented measures to increase child and youth participation in planning (http://www.ucdenver.edu/academics/colleges/ArchitecturePlanning/discover/centers/CYE/Projects/Child-FriendlyCities/ChildYouthFriendlyDenver/Pages/ChildYouthFriendlyDenver.aspx), while the American Planning Association offers guidance on creating family friendly communities (Warner 2008). Carlson (2005) cites the example of Hampton, a city of 148,000 inhabitants in the state of Virginia that has employed young people in its planning department since 1996. Based on her experiences with this project, she concludes that 'engaging young people in government and the civic life of their community truly increases the social capital of the community' (ibid.: 224) and that 'a system of youth civic engagement improves the city's decisions, and better decisions improve the city' (ibid.: 225).

Box 6.6 **The promotion of children's participation in Italy**

In Italy, children and adolescents are included in public decision making through children's councils, young people's conferences, discussion fora and participatory planning projects. The city of Fano created a children's town council in 1992. Members of the children's council develop planning proposals and submit annual requests to the city's municipal council. Examples of measures requested by the children's council include: closing the main streets to traffic for events one day per year; the use of sports installations without having to belong to a sports association; the use of the town squares as places to play; and the use of the CSDBB award to convert a cottage for play and educational activities for children. The children's council also developed a Pedestrians' Charter in response to mobility problems in the city and initiated inter-generational workshops.

Children and adolescents further contribute to planning decisions through participatory project workshops which, in consultation with a planner and an architect, focus on a specific issue and urban area each year, for example piazzas and monuments, green space, traffic, garbage disposal, school and play space. The workshops have covered the whole spectrum of environmental matters, including eco-management of urban waste. Projects have included the conversion of a cottage, the rehabilitation of an abandoned green area, and the creation of a cycle path, a fountain in a courtyard and a sea view point in a park.

Source: Adapted from Corsi (2002: 172–174)

Conclusion

Planning *with* rather than *for* children has become increasingly accepted as an essential approach to creating cities that are not only good *for* children, but also *according to* and *by* children (Riggio 2002). In this chapter we have demonstrated the benefits of participatory planning for young people as well as for wider communities, planners and political decision makers. We have argued that children and youth are experts of their own lifeworlds and need to be consulted about projects that affect them and their community. Participation not only enables planning that is more responsive to the needs of young people. It also builds respect, increases integration and feelings of belonging, allows the development of new competencies and contributes to learning democratic citizenship skills. In order to make participation possible for a wide range of young people and not just for a select few, choosing a wide range of methods is important. This allows young people with different skills, abilities and interests to become engaged in planning projects. Effective participation is often hindered, however, by a number of problems. Thus, adults may hold negative assumptions about young people and find it difficult to let go of positions of power. Consulting children and youth can be tokenistic, decorative or manipulative, leading to little dialogue and interaction. Decisions are often still taken behind closed doors and bureaucratic 'red-tape' as well as institutional structures may mean that results take too long to materialise. Finally, many projects are one-off events that lack sustainability and long-term effects.

Despite these concerns, major strides have been made to include young people more effectively in urban planning. Responding to the UN Convention on the Rights of the Child, municipal authorities of signatory countries around the world have implemented strategies to incorporate the principles of the UNCRC into their own governance structures. We have included a number of examples in this chapter which we hope will be informative, thought provoking and inspiring. Adopting some of the methods and principles outlined here means not being able to predict the outcomes, but to contribute to the reinvention of cities with and for young people. For it is only *with* children and youth that we can imagine what a truly child-friendly city will look like.

Questions for discussion

1. What are the challenges and benefits of conducting participatory planning with young people?
2. How can young people's citizenship be enhanced through participatory planning and research?
3. Which obstacles stand in the way of developing child-friendly cities?

Suggested reading

Chawla, L. (ed.) (2002a) *Growing Up in an Urbanising World*. London: Earthscan.

Driskell, D. (2002) *Creating Better Cities with Children and Youth. A Manual for Participation*. London: Earthscan.

Freeman, C., Henderson, P. and Kettle, J. (1999) *Planning with Children for Better Communities. The Challenge to Professionals*. Bristol: The Policy Press (Community Development Foundation).

Hart, R. (1997) *Children's Participation. The Theory and Practice of Involving Young Citizens in Community Development and Environmental Care*. London: Earthscan.

Journals

Environment and Urbanization 2002, 14, special issue on 'Child-Friendly Cities'.

Children, Youth and Environment 2006, 16, special issue on 'Participatory Research and Children and Youth'.

Useful websites

British Youth Council: http://www.byc.org.uk/

European Youth Forum: http://www.youthforum.org/

Child Friendly Cities: http://childfriendlycities.org/

Child and youth participation: http://www.unicef.org/adolescence/index_38074.html

7 Conclusion

In this chapter we will:
- **summarise the main themes and arguments of the book**
- **outline developments that are likely to affect the lives of urban children and youth in the future.**

> Children mostly live their lives within the warps and wefts of the striations of adult space. These material, symbolic and disciplinary structures are both incidental and deliberate in their relation to children. Children's geographies operate within these patterns. The question is the nature of the interaction between the two. If adults' geographies are intensive, rigid and powerfully embedded, there may be little chance for children to build their own geographies, but if adults' geographies can be more permeable, heterogeneous and tolerant of otherness, then those in society most celebrated for their bodily and mental spontaneity, creativity, exuberance and mobility, may have the chance to express this in the creation of their own geographies within the adult world which, it seems, is bound to continue to be the dominant ordering of space.
>
> (Jones 2000: 43f)

Young people's lives in the city are strongly shaped by adult priorities and structures, yet cities are also, at least in part, the result of their spatial practices and of interactions between adults, children and youth. In this book we have sought to show the extent to which cities are *of* children and youth, and the degree to which urban childhood and youth are influenced by the diverse patterning and texturing of space in different cities around the globe. We have shown how social constructions of age and the city interrelate to produce discourses of belonging and exclusion from (parts of) the city that are based on understandings of childhood and youth in relation to other age groups, yet are context specific and culturally variable. In Chapter 2, we examined particularly how discourses of young people *at* risk and young people *as* risk intersect to produce landscapes of fear that lead to the marginalisation of children and youth in the city. We also considered alternative representations and looked at examples of young people's own participation in the production of discourses about people and place. The chapter further engaged with recent thinking in geography and the social sciences, advocating a relational approach to age and attention to extra-discursive, non-representational aspects of childhood and youth.

Chapter 3 focused on the effects of structural disadvantages and social inequalities on children and youth in different cities around the world. We considered not only how and why young people are, as a group, amongst the most disadvantaged of urban residents, who are often on the front-line of cuts in welfare spending, but also showed to what extent gender, race, ethnicity and class intersect with (young) age to produce highly uneven landscapes of childhood and youth. In relation to urban young people in the Global South, we particularly examined the effects of rural to urban migration and the consequences of poverty and poor health for young people's daily practices, including their participation in waged and unwaged work.

Chapters 4 and 5 considered in greater depth how young people participate in the production of urban space through socio-cultural and spatial practices. Again, in Chapter 4, we were careful to pay attention to the different ways in which children and youth with diverse social and cultural identities interact with the urban, highlighting the extent to which processes and practices of 'growing up' in the city reflect cultural understandings, and lived inequalities, gender, sexuality, race, ethnicity, class, (dis)ability and location. This was explored through the lenses of living and playing in the city. Chapter 5 related these insights to the question of how young people engage with different aspects of globalisation processes, showing the extent to which they partake in these processes, while being marginalised by unequal access to leisure and consumer facilities. We further looked, in this chapter, at ways of contesting and challenging both the age-based exclusion and marginalisation of young people from central urban space through subcultural practices.

In keeping with the overall argument of this book that children and youth are active participants in the making of cities and urban spaces, Chapter 6 examined different ways in which young people can influence decision making about the design of cities more effectively. We advocated the greater use of participatory methods and approaches in planning as well as research, in order to challenge age-based power relations that otherwise place young people in positions of marginality and inferiority from the outset. Our main argument here is that child-friendly cities cannot be created without the active involvement of children and youth themselves. In Chapter 6, we gave numerous examples of participatory techniques and projects, but we also engaged critically with some of the challenges involved in applying them effectively. Finally, we asked what some of the key features of child-friendly cities might be.

Issues for the future

We now want to sketch some of the developments that are likely to impact on how children and youth grow up in cities in the near future. The three issues which we focus on here all have roots in the present, but appear likely to have longer-term consequences for young people's urban lives. We are particularly concerned with:

1 the effects of increased surveillance and fear in the splintering city
2 the contradictory dynamics of growing and shrinking cities
3 the effects of welfare cuts in the aftermath of global recession.

Critical urban theory has, for some time, raised attention to the reduction of public, shared spaces in the city and the tendency to deal with the tensions of urban diversity by increasing segregation rather than interaction and possibilities for encounter (Sennett 2005; Low and Smith 2006; Low 2001). These developments have occurred partly in response to increasing urban complexity and partly as a result of greater wealth divisions under neoliberal regimes of welfare cuts and governmental non-intervention. The paring down of public spending on communal infrastructures has been accompanied by 'zero-tolerance' policies that legitimise the purification of urban space as a domain for middle-class consumption. Increased surveillance and militarisation has entered both our urban centres and the residential enclaves of gated communities:

> The municipally-controlled street systems, that once acted as effective monopolies of the public realm in many cities, are being paralleled by the growth of a set of shadow, privatised street spaces. In these, access is carefully monitored and controlled, often through the use of the latest, Gulf War-standard surveillance technology. The new enclosed street spaces of shopping malls are everywhere built and controlled by private consortia. Privately-run mall spaces are taking over larger and larger parts of towns and city centres. Private security and closed circuit (CCTV) operating companies are being given responsibility for maintaining public order and deciding who can, and who cannot, go in and out of such spaces unhindered and unscrutinised. And domestic street spaces, too, are changing, as gated communities are growing fast in virtually all urban contexts around the world.

> A key to all such privatised and splintered street spaces is that they abandon the principle of free, open and democratic access in favour of a policy of actively restraining and excluding those deemed not to belong.
>
> (Graham 2001: 365)

These new urban landscapes of surveillance (see Chapter 4) are legitimised by a discourse of fear, which only too often focuses on young people as particularly deviant segments of society (Giroux 2003), as we have shown in Chapter 2. Young people are demonised *per se* and governmental crime reduction schemes target them particularly through measures such as restraining and antisocial behaviour orders (Woolley 2007; France 2008). These measures take precedence over strategies to tackle poverty and social exclusion and instead individualise the causes of delinquency (France 2008).

While fear of urban others, including youth, is not in itself a new phenomenon and the policies just described have been a long time in the making, what it is important to recognise is their increasing connection to a broader geopolitics of fear following the events of 11 September 2001 (Pain and Smith 2008). The latter gives further justification to, and leads to, an intensification of surveillance that continues to limit young people's access to public space. It also reduces further the potential of the city to provide space for social interaction and the negotiation of difference, increasing divisions between diverse social and cultural groups and the marginalisation of those marked as 'other' in the moral panic ignited by the 'war on terror'. Henri Giroux, in his critical analysis of the effects of post-9/11 policy making for young people, argues that '[y]outh have become the all important group onto which class and racial anxieties are projected' (2003: xvi). He strategically reverses the gaze, however, to highlight the extent to which children and youth have themselves become the targets of a war waged on them by attacks on children's rights, social services, the welfare state and public schools, political developments which affect particularly those marginalised by class and colour (ibid). Giroux thus focuses attention on the issue of social disinvestment, which is a far greater threat for individual children and youth than the risks associated with terrorism.

Disinvestment is likely to be a key challenge for the welfare of young people in both growing and shrinking cities. Many cities in the Global South have experienced significant growth over the last three decades as a result of rural–urban migration and population increases (United Nations Center for Human Settlements 1996; Drakakis-Smith 2000). Investments in affordable, safe housing and in the infrastructures necessary for achieving even a minimum standard of living have not kept up with these levels of growth, partly because of the weakness of local economies and the uneven global distribution of wealth.

Many of the examples we discuss in Chapters 3 and 4 demonstrate how this lack of investment is affecting the lives of children and youth in these cities, while Chapter 6 cites case studies of the child-friendly cities movement which seeks to develop remedial measures to some of the worst effects of disinvestment for children. Environmental degradation and rural deprivation are likely to continue to fuel migration to the cities in the future, but population growth is slowing down due to poor health and particularly the AIDS pandemic (Lewis *et al.* 2004; Ansell and van Blerk 2004a). Unless there is greater global commitment to tackle AIDS, for example through the provision of cheaper medication, the average age of people in the worst affected and poorest nations of the world will remain low, meaning high numbers of orphaned children who are having to take on adult responsibilities from an ever younger age without sufficient support and access to quality education and health services.

Insufficient investment in public services and infrastructures is a problem not only for growing cities in the Global South, but also for those former urban centres on the margins of post-industrial society which have lost their economic basis as a result of global economic and/or national political change. This includes cities in the former manufacturing belts of North America and Western Europe, only some of which have managed to gain a new foothold in service and high-tech industries, but it particularly affects cities in parts of the former socialist world, which have lost their moorings in the socialist economic network without gaining sufficient opportunities for sustainable growth in the global economy. These 'shrinking' cities and regions (Oswalt 2004; Glaeser and Gyourko 2005) are suffering significant population loss due to outmigration, which is particularly pronounced amongst skilled and well-educated younger people. Urban planners and policy makers as a result focus their priorities not only on providing services for an ageing population, but also on planning for reduced public needs. Their opportunities for investment in new public infrastructures are severely restrained and oriented towards a potentially unsustainable future. This has consequences not only for an ageing population relying on constrained public resources and fewer people of working age to provide services and sustain the local economy, but also for younger residents who are not able, needing or wanting to migrate, since they are faced with fewer resources targeted at their needs. In eastern Germany, this has meant rising costs for childcare and a significant reduction in the numbers of schools, youth centres and nurseries, with less choice for parents and their children and the need to travel greater distances for basic public services. 'Shrinking' cities thus present their own set of challenges for maintaining and improving welfare standards for children, especially as families also suffer most from declining income levels and rising levels of poverty (Förster 2004).

Many of the problematic issues that arise from both growing and shrinking cities in terms of resource shortages and uneven access to urban facilities for children and youth are propounded by the current intensification and entrenchment of social inequalities in the aftermath of global recession. Governments around the world are introducing welfare cuts that will, almost without exception, hit those hardest who are already in the most difficult socio-economic position. Families with children, particularly single-parent households, those with many children and those households headed by children will inevitably be amongst the hardest hit as spending on development aid and welfare budgets is cut. They carry the double burden of care not just for themselves but for others in the wider networks of social reproduction, meaning that the social inequalities produced today will carry on to affect the life chances of tomorrow's adults and their children, as well as the quality of life of others who depend on their work and care. Instead of producing nurturing environments for children and youth from diverse backgrounds, cities are likely to become characterised by further wealth division and more, rather than less, securitisation in order to 'contain' urban unrest.

The three issues which we have highlighted here are, of course, only some of the influences that are likely to bear on young people's urban lives over the next decade or so. Whether they will have the effects we anticipate will depend to a large degree on the commitment of future governments to place the well-being of children and youth at the heart of their policies and to cooperate globally to alleviate the vast inequalities in young people's living standards across the world. Positive outcomes will also depend, however, on the willingness of adult decision makers to listen to the voices of children and youth and, more importantly, to look in earnest for ways to implement their ideas and respond to the interests and needs they articulate. As Matthews and Limb suggest in their 'agenda for a geography of children' (1999), children need allies, because they do not have the same access to academic institutions and political decision-making structures as adults:

> unlike many other groups who are marginalized as 'outsiders' within society (for example, women, the disabled, ethnic minorities) children occupy a special position of exclusion for they will never gain entry to the academy. (The academy is used here as a collective term to describe sets of hegemonic values which form part of the apparatus of ruling.) Their ability to challenge the conventions of dominant ideology from within, together with the practices and processes which lead to their marginalization, is mostly beyond their grasp. Children need allies. For these allies there remains to be solved the contradiction between the world from a child's viewpoint and the world they experience as adults.
>
> (ibid.: 83)

We hope that this book has not only provided relevant insights for the study of children, youth and the city, but also inspired reflection on the validity of current planning and policy priorities and on ways in which young people's own priorities can be made more central to the design of cities.

References

Abrams, M. (1959) *The Teenage Consumer*, London: London Press Exchange.

Abu-Ghazzeh, T. (1998) 'Children's use of the street as a playground in Abu-Nuseir, Jordan', *Environment and Behaviour* 30(6): 799–831.

Abu-Lughod, J. (1961) 'Migrant adjustment to city life: the Egyptian case', *The American Journal of Sociology*, 67(1): 22–32.

Adams, E. and Ingham, S. (1998) *Changing Places. Children's Participation in Environmental Planning*, London: The Children's Society.

Agamben, G. (2005) *States of Exception*, Chicago: Chicago University Press.

Aguirre, Jr, A., Eick, V. and Reese, E. (eds) (2006) 'Introduction: neoliberal globalization, urban privatization, and resistance', *Social Justice* 33(3): 1–5.

Ahmend, A. and Sohail, M. (2008) 'Child's play and recreation in Dhaka City, Bangladesh', *Proceedings of the Institution of Civil Engineers: Municipal Engineer* 161(ME4): 263–270.

Aitken, S. (2001) *Geographies of Young People: The Morally Contested Spaces of Identity*, London: Routledge.

Alanen, L. (2000) 'Childhood as generational condition: towards a relational theory of childhood in research', in *Childhood: Sociology, Culture and History. A Collection of Papers*, Odense: University of Southern Denmark.

Alanen, L. (2001) 'Explorations in generational analyses', in Alanen, L. and Mayall, B. (eds) *Conceptualizing Child-Adult Relations*, London: Routledge, pp. 11–22.

Aldrich, R. (2004) 'Homosexuality and the city: an historical overview', *Urban Studies* 41(9): 1719–1737.

Alexander, C. (2009) 'Deviant femininities: the everyday making and unmaking of "criminal" youth', in Hörschelmann, K. and Colls, R. (eds) *Contested Bodies of Childhood and Youth*, London: Palgrave Macmillan.

Alexandrescu, G. (1996) 'Programme note: street children in Bucharest', *Childhood: A Global Journal of Child Research* 3(2): 267–270.

Alvarez-Rivadulla, M. (2007) 'Golden ghettos: gated communities and class residential segregation in Montevideo, Uruguay', *Environment and Planning A* 39: 47–63.

Amin, A. and Thrift, N. (2002) *Cities: Reimagining the Urban*, Cambridge: Polity Press.

Amit-Talai, V. and Wulff, H. (1995) *Youth Cultures: A Cross-Cultural Perspective*, London: Routledge.

Andersson, J. (2010) 'Vauxhall's post-industrial pleasure gardens: "Death wish" and hedonism in 21st-century London', *Urban Studies* 48(1): 85–100.

Andreasen, J. (1996) *Urban Tenants and Community Involvement.* Habitat International 20(3): 359–365.

Ansell, N. (2005) *Children, Youth and Development*, London: Routledge.

Ansell, N. and van Blerk, L. (2004a) 'Children's migration as a household/family strategy: coping with AIDS in southern Africa', *Journal of Southern African Studies* 30(3): 673–690.

Ansell, N. and van Blerk, L. (2004b) *HIV/AIDS and Children's Migration in Southern Africa*, SAMP Migration Policy Series No. 33, Cape Town: SAMP.

Ansell, N. and van Blerk, L. (2005) '"Where we stayed was very bad ...": migrant children's perspectives on life in informal rented accommodation in two southern African cities', *Environment and Planning A* 37: 423–440.

Appadurai, A. (1996) *Modernity at Large: Cultural Dimensions of Globalization*, Minneapolis: University of Minnesota Press.

Appold, S. and Yuen, B. (2007) 'Families in flats, revisited', *Urban Studies* 44(3): 569–589.

Aries, P. (1962) *Centuries of Childhood*, New York: Vintage Press.

Ataöv, A. and Haider, J. (2006) 'From participation to empowerment: critical reflections on a participatory action research project with street children in Turkey', *Children, Youth and Environments* 16(2): 127–152.

Attree, P. (2006) 'The social costs of child poverty: a systematic review of the qualitative evidence', *Children and Society* 20: 54–66.

Austin, J. (2001) *Taking the Train. How Graffiti Art Became an Urban Crisis in New York City*, New York: Columbia University Press.

Bailey, A. (2009) 'Population geography: lifecourse matters', *Progress in Human Geography*, 33(3): 407–418.

Bain, A. (2003) 'White Western teenage girls and urban space: challenging Hollywood's representations', *Gender, Place and Culture* 10: 197–213.

Bannerjee, K. and Driskell, D. (2002) 'Tales from Truth Town. Children's lives in a south Indian "slum"', in Chawla, L. (ed) *Growing Up in an Urbanising World*, London: Earthscan, pp. 135–160.

Bar-On, A. (1997) 'Criminalising survival: images and reality of street children', *Journal of Social Policy* 26(1): 63–78.

Baraldini, C. (2003) 'Planning childhood: children's social participation in the town of adults', in Christensen, P. and O'Brien, M. (eds) *Children in the City. Home, Neighbourhood and Community*, London: Routledge, pp. 184–205.

Barker, J. and Weller, S. (2003) 'Never work with children? Methodological issues in children's geographies', *Qualitative Research* 3(2): 207–227.

Bartlett, S. (1999) 'Children's experience of the physical environment in poor urban settlements and the implications for policy, planning and practice', *Environment and Urbanization* 11: 63–74.

Bartlett, S., Hart, R., Satterthwaite, D., De La Barra, X. and Missair, A. (1999) *Cities for Children: Children's Rights, Poverty and Urban Management*, London: Earthscan.

Bauman, Z. (1998) *Globalization. The Human Consequences*, Oxford: Polity Press.

Bavidge, J. (2006) 'Stories in space: The geographies of children's literature', *Children's Geographies* 4(3): 319–330.

BBC News (2006) 'Asbos 'demonising' young people', http://news.bbc.co.uk/go/pr/fr/-/1/hi/uk/4935606.stm 23/04/2006.

Beazley, H. (2000) 'Home sweet home?: Street children's sites of belonging', in Holloway, S. and Valentine, G. (eds), *Children's Geographies: Playing, Living, Learning,* London: Routledge, pp. 194–212.

Beazley, H. (2002) '"Vagrants wearing make-up": negotiating spaces on the streets of Yogyakarta', *Urban Studies* 39(9): 1665–1684.

Beazley, H. (2003) 'Voices from the margins: street children's subcultures in Indonesia', *Children's Geographies* 1(2): 181–200.

Beazley, H. (2008) '"I Love Dugem": young women's participation in the Indonesian dance scene', in Bennett, L. and Parker, L. (eds) *Intersections: Gender and Sexuality in Asia and the Pacific*, 18, November, http://intersections.anu.edu.au/issue18_contents.htm.

Beazley, H. and Chakraborty, K. (2008) 'Cool consumption: Rasta Punk and Bollywood on the streets of Yogyakarta, Indonesia and Kolkata, India', in Rodrigues, U. M. and Smaill, B. (eds) *Youth, Media and Culture in the Asia Pacific Region*, Newcastle: Cambridge Scholars Publishing, pp. 195–214.

Beazley, H., Bessell, S., Ennew, J. and Waterson, R. (2009) 'The right to be properly researched: research with children in a messy, real world', *Children's Geographies* 7(4): 365–378.

Beck, U. and Beck-Gernsheim, E. (2002) *Individualization: Institutionalized Individualism and Its Social and Political Consequences*, London: Sage.

Bell, D. and Valentine, G. (1995) *Mapping Desire*, London: Routledge.

Bell, D., Binnie, J., Cream, J. and Valentine, G. (1994) 'All hyped up and no place to go', *Gender, Place and Culture* 1: 31–48.

Benwell, M. (2009) 'Challenging minority world privilege: children's outdoor mobilities in post-apartheid South Africa', *Mobilities* 4(1): 77–101.

Bennett, A. (1999) *Popular Music and Youth Culture: Music, Identity, and Place*, London: Palgrave.

Bennett, A. (2000) *Popular Music and Youth Culture: Music, Identity and Place*, Basingstoke: Macmillan.

Berglund, U. and Nordin, K. (2007) 'Using GIS to make young people's voices heard in urban planning', *Built Environment* 33(4): 469–481.

Bingham, N., Holloway, S. L. and Valentine, G. (1999) 'Where do you want to go tomorrow? Connecting children and the internet', *Environment and Planning D: Society and Space* 17: 655–672.

Bingley, A. and Milligan, C. (2007) '"Sandplay, clay and sticks": multi-sensory research methods to explore the long-term mental health effects of childhood play experience', *Children's Geographies* 5(3): 283–296.

Bondi, L. (1998) 'Gender, class and urban space: public and private space in contemporary urban landscapes', *Urban Geography* 19(2): 160–185.

Bondi, L. and Rose, G. (2003) 'Constructing gender, constructing the urban: a review of Anglo-American feminist urban geography', *Gender, Place and Culture* 10(3): 229–245.

Borden, I. (2001) *Skateboarding, Space and the City. Architecture and the City*, Oxford: Berg.

Bourdieu, P. (1977) *Outline of a Theory of Practice*, Cambridge: Cambridge University Press.

Bourdieu, P. (1984) *Distinction: A Social Critique of the Judgement of Taste*, London: Routledge.

Bowlby S., Lloyd Evans, S. and Mohammad, R (1998) Becoming a paid worker: Images and identity in Skelton and Valentine (eds) Cool Places: geographies of youth cultures, London: Routledge, pp. 229–248.

Boyden, J. and Ennew, J. (1997) *A Manual for Participatory Research with Children*, Sweden: Radda Barnen.

Bradshaw, J. (2002) 'Child poverty and child outcomes', *Children and Society* 16: 131–140.

Bradshaw, J. (2003) 'Poor children', *Children and Society* 17: 162–172.

Brenner, N. and Theodore, N. (eds) (2002) 'Cities and the geographies of "actually existing neoliberalism", in *Spaces of Neoliberalism: Urban Restructuring in North America and Western Europe*, Malden, MA: Oxford's Blackwell Press, pp. 2–32.

Bridge, G. (2006) 'It's not just a question of taste: gentrification, the neighbourhood, and cultural capital', *Environment and Planning A* 38: 1965–1978.

Brooke, M. (1999) *The Concrete Wave. The History of Skateboarding*, Toronto LA: Warwick Publishing.

Butcher, M. and Velayutham, S. (eds) (2009) *Dissent and Cultural Resistance in Asia's Cities*, London: Routledge.

Butler, J. (1989) *Gender Trouble: Feminism and the Subversion of Identity*, London: Routledge.

Butler, J. (1993) *Bodies That Matter: On the Discursive Limits of 'Sex'*, London: Routledge.

Cahill, C. (2004) 'Defying gravity? Raising consciousness through collective research', *Children's Geographies* 2(2): 273–286.

Camacho, A. (1999) 'Family, child labour and migration: child domestic workers in Metro Manila', *Childhood* 6(1): 57–73.

Cameron, L. (2006) 'Science, nature and hatred: 'finding out' at the Malting House garden School, 1924–29', *Environment and Planning D: Society and Space* 24: 851–872.

Campbell, D. (2005) 'Revealed: Britain's network of child drug runners', in *The Observer*, 16 October 2005, http://observer.guardian.co.uk/uk_news/story/0,6903,1593312,00.html (accessed 6 July 2007).

Camstra, R. (ed.) (1997) *Growing Up in a Changing Landscape*, Assen: Van Gorcum.

Carlson, C. (2005) 'Youth with influence: the youth planner initiative in Hampton, Virginia', *Children, Youth and Environments* 15(2): 211–226.

Carter, R. (2000) *The silent crisis: the impact of poverty on children in Eastern Europe and the Former Soviet Union*, London: European Children's Trust.

Castells, M. (2000) *End of Millenium*, Malden: Blackwell.

Certeau, M. de (1984) *The Practice of Everyday Life*, Berkeley: University of California Press.

Chatterton, P. and Hollands, R. (2001) Theorising urban playscapes: producing, regulating and consuming youthful nightlife city spaces. *Urban Studies* 39(1): 95 116.

Chatterton, P. and Hollands, R. (2002) 'Theorising urban playspaces: producing, regulating and consuming youthful nightlife city spaces', *Urban Studies* 39(1): 95–116.

Chatterton, P. and Hollands, R. (2003) *Urban Nightscapes. Youth Cultures, Pleasure Spaces and Corporate Power*, London: Routledge.

Chau, K. W., Wong, S. K., Yau, Y. and Yeung, A. K. C. (2007) 'Determining optimal building height', *Urban Studies* 44(3): 591–607.

Chawla, L. (1999) 'Life paths into effective environmental action', *Journal of Environmental Education* 31(1): 15–26.

Chawla, L. (ed.) (2002a) *Growing Up in an Urbanising World*, London: Earthscan.

Chawla, L. (2002b) 'Introduction. Cities for human development', in Chawla, L. (ed.) *Growing Up in an Urbanising World*, London: Earthscan, pp. 15–34.

Chawla, L. (2002c) 'Conclusion. Toward better cities for children and youth', in Chawla, L. (ed.) *Growing Up in an Urbanising World*, London: Earthscan, pp. 219–242.

Chawla, L. (2002d) 'Insight, creativity and thoughts on the environment": integrating children and youth into human settlement development', *Environment and Urbanization* 14(2): 11–22.

Chawla, L. and Heft, H. (2002) 'Children's competence and the ecology of communities: a functional approach to the evaluation of participation', *Journal of Environmental Psychology* 22: 201–216.

Chawla, L. and Malone, K. (2003) 'Neighbourhood quality in children's eyes', in Christensen, P. and O'Brien, M. (eds) *Children in the City. Home, Neighbourhood and Community*, London: Routledge, pp. 118–141.

Chin, E. (2001) *Purchasing Power: Black Kids and American Consumer Culture*, Minneapolis: University of Minneapolis Press.

Christensen, P. and James, A. (eds) (2008) *Research with Children: Perspectives and Practices*, New York and London: Routledge.

CINI-ASHA (2003) 'Family adjustments for mainstreaming child labourers into formal schools in Calcutta: the experience of CINI-ASHA', in Kabeer, N., Nambissan, G. B. and Subrahmanian, R. (eds) *Child Labour and the Right to Education in South Asia: Needs Versus Rights?*, London: Sage.

Clark, A. and Percy-Smith, B. (2006) 'Beyond consultation: participatory practices in everyday spaces', *Children, Youth and Environments* 16(2): 1–9.

Clarke, J., Hall, S., Jefferson, T. and Roberts, B. (1976) *Resistance through Ritual. Youth Subcultures in Post-War Britain*, London: Hutchinson.

Cohen, S. (1972) *Folk Devils and Moral Panics*, London: MacGibbon and Kee.

Cohen, R. and Long, G. (1998) 'Children and anti-poverty strategies', *Children and Society* 12: 73–85.

Coleman, J. S. (1961) *The Adolescent Society*, New York: Free Press.

Collins, M. (1997) *Altered State: The Story of Ecstacy Culture and Acid House*, London: Serpent's tail.

Colls, R. and Hörschelmann, K. (2009) 'Introduction: embodied geographies of childhood and youth', *Children's Geographies*, with Rachel Colls, 7(1): 1–6.

Connolly, M. and Ennew, J. (1996) 'Introduction: children out of place', *Childhood: A Global Journal of Child Research,* children out of place: special issue on working and street children, 3(2): 131–145.

Connolly, P. (1998) *Racism, Gender Identities and Young Children: Social Relations in a Multi-Ethnic, Inner-City Primary School*, London: Routledge.

Conticini, A. (2005) 'Urban livelihoods from children's perspectives: protecting and promoting assets on the streets of Dhaka', *Environment and Urbanisation* 17(2): 69–81.

Conticini, A. and Hulme, D. (2007) 'Escaping violence, seeking freedom: why children in Banglasdesh migrate to the street', *Development and Change* 38(2): 210–227.

Corsi, M. (2002) 'The child friendly cities initiative in Italy', *Environment and Urbanization* 14(2): 169–180.

Cosco, N. (2002) 'Extending the Participatory Process to Children with Special Needs' in Driskell, D. (ed.) *Creating Better Cities with Children and Youth: A Manual for Participation*, London: Earthscan/UNESCO.

Cox, R. and Narula, R. (2003) Playing Happy Families: Rules and relationships in au pair employing households in London, England, *Gender, Place and Culture*, 10(4): 333–344.

Cresswell, T. (1992) 'The Crucial "Where" of Graffiti: a geographical analysis of reactions to graffiti in New York', *Environment and Planning D: Society and Space* 10: 329–344.

Cunningham, H. (1991) *The Children of the Poor: Representations of Childhood since the Seventeenth Century*, Oxford: Blackwell.

Cunningham, C., Jones, M. A. and Dillon, R. (2003) 'Children and urban regional planning: participation in the public consultation process through story writing', *Children's Geographies* 1(2): 201–221.

Dallape, F. (1996) 'Urban children: a challenge and an opportunity', *Childhood: A Global Journal of Child Research,* children out of place: special issue on working and street children, 3(2): 283–294.

Dalsgaard, A., Franch, M. and Parry Scott, R. (2008) 'Dominant ideas, uncertain lives: the meaning of youth in Recife', in Hansen, K. (ed.) (2008) *Youth and the City in the Global South*, Bloomington: Indiana University Press, pp.49–73.

Davin, A. (1996) *Growing Up Poor: Home, School and Street in London 1870–1914,* London: Rivers Oram Press.

Davis, M. (1992) 'Fortress Los Angeles. The militarisation of urban space', in Sorkin, M. (ed.) *Variations on a Theme Park*, New York: Noonday Press, pp. 154–180.

Delap, E. (2000) 'Urban children's work during and after the 1998 floods in Bangladesh', *Development in Practice* 10(5): 662–673.

Delap, E. (2001) 'Economic and cultural forces in the child labour debate: evidence from urban Bangladesh', *The Journal of Development Studies* 37(4): 1–22.

Dennis, R. (1997) 'Property and propriety: Jewish landlords in early twentieth-century Toronto', *Transactions of the Institute of British Geographers, New Series*, 22(3): 377–397.

De Visscher, S. and Bouverne-de Bie, M. (2008) 'Recognising urban public space as a co-educator: children's socialisation in Ghent', *International Journal of Urban and Regional Research* 32(3): 604–616.

Dodman, D. (2003) 'Shooting in the city: an autophotographic exploration of the urban environment in Kingston, Jamaica', *Area* 35(3): 293–304.

Douglas, H. (2006) 'Action, blastoff, chaos: ABCs of successful youth participation', *Children, Youth and Environments* 16(2): 347–365.

Drakakis-Smith, D. (2000) *Third World Cities*, London: Routledge.

Driskell, D. (2002) *Creating Better Cities with Children and Youth. A Manual for Participation*, London: Earthscan.

D'Souza, R. (1997) Housing and environmental factors and their effects on the health of children in the slums of Karachi, Pakistan, *Journal of Biosocial Science*, 29: 271–281.

Dürrschmidt, J. (1997) 'The delinking of locale and milieu. On the situatedness of extended milieux in a global environment', in Eade, J. (ed.) *Living in the Global City. Globalization as Local Process*, London: Routledge, pp. 56–72.

Dyer, C (2007) 'Working children and educational inclusion in Yemen', *International Journal of Educational Development* 27(5): 512–524.

Eade, J. (ed.) (1997) *Living in the Global City. Globalization as Local Process*, London: Routledge.

Elden, S. (2004) *Understanding Henri Lefebvre. Theory and the Possible*, New York: Continuum.

Elden, S., Lebas, E. and Kofman, E. (2003) *Henri Lefebvre. Key Writings*, New York: Continuum.

Elder, G. H. (1994) 'Time, human agency, and social change: perspectives on the life course', *Social Psychology Quarterly* 57: 4–15.

Elsley, S. (2004) 'Children's experience of public space', *Children and Society* 18: 155–164.

Englund, H. (2002) 'The village in the city: the city in the village: migrants in Lilongwe', *Journal of Southern African Studies* 28: 137–154.

Ennew, C. (1994) ‘Time for children or time for adults?’, in Qvortrup, J., Bardy, M., Sgritta, G. and Wintersberger, H. (eds) *Childhood Matters: Social Theory, Practice and Politics*, Aldershot: Avebury, pp. 125–143.

Epstein, J. S. (1998) ‘Introduction: Generation X, youth culture, and identity’, in Epstein, J. S. (ed.) *Youth Culture. Identity in a Postmodern World*, Oxford: Blackwell, pp. 1–23.

Erikson, E. (1968) *Identity: Youth and Crisis*, London: Faber.

Eubanks-Owens, P. (1997) ‘Adolescence and the Cultural Landscape: Public Policy, Design Decisions and Popular Press Reporting’, *Landscape & Urban Planning* 39: 153–166.

Evans, R. (2005) ‘Social networks, migration and care in Tanzania: caregivers’ and children’s resilience in coping with HIV/ AIDS’, *Journal of Children and Poverty*, 11(2) 111–129.

Evans, R. (2011) ‘Young caregiving and HIV in the UK: caring relationships and mobilities in African migrant families’, *Population, Space and Place* 17(4): 338–360.

Evans, R. and Becker, S. (2011) *Children Caring for Parents with HIV and AIDS: Global Issues and Policy Responses*, Bristol: Policy Press.

Eversole, H. (2009) ‘Asking children: the benefits from a programming perspective’, *Children’s Geographies* 7(4): 484–486.

Farrer, J. (1999) ‘Disco “super-culture”: consuming foreign sex in the Chinese disco: cosmopolitan dance culture and cosmopolitan sexual culture’, *Sexualities* 2: 147–166.

Farrer, J. (2002) *Opening Up: Youth Sex Culture and Market Reform in Shanghai*, Chicago: University of Chicago Press.

Featherstone, M. (1995) *Undoing Culture. Globalization, Postmodernism and Identity*, London: Sage.

Ferrell, J. (1996) *Crimes of Style. Urban Graffiti and the Politics of Criminality*, Boston, MA: Northeastern University Press.

Fincher, R. and Jacobs, J. (1998) *Cities of Difference*, London: Guildford Press.

Fisher, C. S. (1976) *The Urban Experience*, New York: Harcourt Brace Jovanivich.

Fitzhugh, L. (1964) *Harriet the Spy*, New York: Harper and Row.

Flaherty, J., Veit-Wilson, J. and Dornan, P. (2004) *Poverty: the Facts*, 5th edn, London: Child Poverty Action Group.

Fontes, M. B., Hillis, J. and Wasek, G. K. (1998) ‘Children affected by AIDS in Brazil: estimates of the number of children at risk of being orphaned and displaced by AIDS in Brazil’, *Childhood* 5(3): 345–363.

Förster, P. (2004) *Ohne Arbeit keine Freiheit. Warum junge Ostdeutsche rund 15 Jahre nach dem Susammenbruch des Sozialismus noch nicht im gegenwärtigen Kapitalismus angekommen sind*, Leipzig: Rosa-Luxemburg-Stiftung.

Foster, J. (1997) *Our Bairns. Glimpses of Tyneside's Children c.1850–1950*, Newcastle: Newcastle Libraries and Information Service.

Foucault, M. (1972) *The Archaeology of Knowledge*, London: Tavistock.

Foucault, M. (1977) *Discipline and Punish*, London: Allen Lane.

Foucault, M. (1980) *Power/Knowledge*, Brighton: Harvester.

Fourchard, L. (2006) 'Lagos and the invention of juvenile delinquency in Nigeria, 1920–60', *Journal of African History* 47: 115–137.

France, A. (2008) 'Risk factor analysis and the youth question', *Journal of Youth Studies* 11(1): 1–15.

Francis, M. and Lorenzo, R. (2002) 'Seven realms of children's participation', *Journal of Environmental Psychology* 22: 157–169.

Freeman, C., Henderson, P. and Kettle, J. (1999) *Planning with Children for Better Communities. The Challenge to Professionals*, Bristol: The Policy Press (Community Development Foundation).

Freeman, C., Nairn, K. and Sligo, J. (2003) '"Professionalising" participation: from rhetoric to practice', *Children's Geographies* 1(1): 53–70.

Gagen, E. (2000a) 'An example to us all: child development and identity construction in early 20th-century playgrounds', *Environment and Planning A* 32: 599–616.

Gagen, E. (2000b) 'Playing the part: performing gender in America's playgrounds', in Holloway, S. and Valentine, G. (eds) *Children's Geographies: Playing, Living, Learning*, London: Routledge.

Gagen, E. (2006) 'Measuring the soul: psychological technologies and the production of physical health in Progressive Era America', *Environment and Planning D: Society and Space* 24: 827–849.

Gagen, E. (2009) 'Commentary: disciplining bodies', in Hörschelmann, K. and Colls, R. (eds) *Contested Bodies of Childhood and Youth*, London: Palgrave Macmillan.

Gelder, K. (1997) 'Introduction to part four', in Gelder, K. and Thornton, S. (eds) *The Subcultures Reader*, London: Routledge, pp. 315–319.

Gelder, K. and Thornton, S. (eds) (1997) *The Subcultures Reader*, London: Routledge.

Giddens, A. (1994) *Modernity and Self-Identity. Self and Society in the Late Modern Age*, Oxford: Polity Press.

Gillespie, S. and Kadiyala, S. (2005) *HIV/AIDS and Food and Nutrition Security: From Evidence to Action*, Food Policy Review 7, Washington DC: International Food Policy Research Institute.

Giroux, H. (2003) *The Abandoned Generation. Democracy Beyond the Culture of Fear*, London: Palgrave Macmillan.

Glaeser, E. L. and Gyourko, J. (2005) 'Urban decline and durable housing', *Journal of Political Economy* 113(2): 345–375.

Goldson, B. (1997) 'Children, crime, policy and practice: neither welfare nor justice', *Children and Society* 11: 77–88.

Goodwin, M. (1995) 'Poverty in the city: "You can raise your voice but who is listening?"', in Philo, C. (ed.) *Off the Map. The Social Geography of Poverty in the UK*, London: Child Poverty Action Group, pp. 65–82.

Gough, K. (2008) 'Youth and the home', in Hansen, K. (ed.) *Youth and the City in the Global South*, Bloomington: Indiana University Press, pp. 127–150.

Gough, I. and McGregor, A. (2007) *Wellbeing in Developing Countries. From Theory to Practice*, Cambridge: Cambridge University Press.

Graham, S (2001) 'The spectre of the splintering metropolis', *Cities* 18(6): 365–368.

Griesel, R., Swart-Kruger, J. and Chalwa, L. (2002) 'Children in South Africa can make a difference: an assessment of "Growing Up in Cities" in Johannesburg', *Childhood* 9(1): 83–100.

Griffin, C. (1993) *Representations of Youth. The Study of Youth and Adolescence in Britain and America*, Cambridge: Polity Press.

Griffin, C. (1997) 'Troubled teens: managing disorders of transition and consumption', *Feminist Review* 55: 4–21.

Griffin, C. (2001) 'Imagining new narratives of youth. Youth research, the 'New Europe' and global youth culture', *Childhood* 8: 147–166.

Guerra, E. (2002) 'Citizenship knows no age: children's participation in the governance and municipal budget of Barra Mansa, Brazil', *Environment and Urbanization* 14(2): 70–84.

Hall, P. (2007) 'The United Kingdom's experience in revitalizing inner cities', in Ingram, G., Hong, Y. (eds) *Land Policies and their Outcomes: Proceedings of the 2006 Land Policy Conference*, Cambridge, MA: Lincoln Institute of Land Policy, pp 259–83.

Hall, S. (1992) 'The West and the rest', in Hall, S. and Gieben, B. (eds) *Formations of Modernity*, Cambridge: Polity Press.

Hall, S. (1997) 'The work of representation', in Hall, S. (ed.) *Representation: Cultural Representations and Signifying Practices*, London: Sage, pp. 13–74.

Hall, S. G. (1904) *Adolescence*, New York: Appleton.

Hall, S. G. (1906) *Youth: Its Education, Regimen and Hygiene*, New York: Appleton.

Hallman, K., Quisumbing, A., Ruel, M. and De La Briere, B. (2005) Mothers' work and child care: findings from the urban slums of Guatemala City, *Economic Development and Cultural Change* pp. 855–885.

Hamnett, C. (1991) 'The blind man and the elephant: the explanation of gentrification', *Transactions of the Institute of British Geographers* 16: 173–189.

Hammett, D. (2010) 'Reworking and resisting globalising influences: Cape Town hip-hop', *GeoJournal*, published online 23 November 2010, DOI: 10.1007/s10708–010–9395–1.

Hannerz, U. (1996) *Transnational Connections. Culture, People, Places*, London: Routledge.

Hareven, T. K. (1995) 'Changing Images of Aging and the Social Construction of the Life Course', in Featherstone, M. and Wernick, A. (eds) *Images of Aging: Cultural Representations of Later Life*, London: Routledge.

Harris, A. (2006) 'Introduction: critical perspectives on child and youth participation in Australia and New Zealand/Aotearoa', *Children, Youth and Environment* 16(2): 220–230.

Hart, R. (1978) *Children's Experience of Place: A Developmental Study*. New York: Irvington Publishers, Inc.

Hart, R. (1992) *Children's Participation: From Tokenism to Citizenship*. Florence: International Child Development Centre, UNICEF.

Hart, R. (1997) *Children's Participation. The Theory and Practice of Involving Young Citizens in Community Development and Environmental Care*, London: Earthscan.

Hart, R. (2002) 'Containing children: some lessons on planning for play in New York City', *Environment and Urbanization* 14(2): 135–148.

Harvey, D. (1973) *Social Justice and the City*, London: Arnold.

Harvey, D. and Chatterjee, L. (1974) 'Absolute rent and the structuring of space by governmental and financial institutions', *Antipode* 6(1): 22–36.

Hassan, R. (1977) *Families in Flats: A Study of Low Income Families in Public Housing*, Singapore: Singapore University Press.

Hebidge, D. (1979) *Subculture: The Meaning of Style*, New York: Methuen.

Hecht, T. (1998) *At Home in the Street*, Cambridge: Cambridge University Press.

Hill, M., Davis, J., Prout, A. and Tisdall, K. (2004) 'Moving the participation agenda forward', *Children and Society* 18: 77–96.

Hodgkin, R. (1998) 'Crime and disorder bill', *Children and Society* 12: 66–68.

Hollands, R. (2002) 'Divisions in the dark: youth cultures, transitions and segmented consumption spaces in the night-time economy', *Journal of Youth Studies* 5(2): 153–171.

Holloway, S. L. and Valentine, G. (2000a) 'Spatiality and the new social studies of childhood', *Sociology* 34: 763–83.

Holloway, S. L. and Valentine, G. (2000b) *Children's Geographies. Playing, Living, Learning*, London: Routledge.

Holloway, S. L. and Valentine, G. (2003) *Cyberkids. Children in the Information Age*, London: Routledge.

Holt, L. (2004) 'Children with mind-body differences: performing disability in primary school classrooms', *Children's Geographies* 2(2): 219–236.

Home Office (2006) *Anti-social behaviour orders*, http://ww.crimereduction.gov.uk/antisocialbehaviour55.htm (accessed 27 September 2006).

Hopkins, P. E. (2006) 'Youthful Muslim masculinities: gender and generational relations', *Transactions of the Institute of British Geographers* 31(3): 337–352.

Hopkins, P. E. (2010) *Young People, Place and Identity*, London: Routledge.

Hopkins, P. and Pain, R. (2007) 'Geographies of age: thinking relationally', *Area* 39: 287–94.

Horelli, L. (1998) 'Creating child-friendly environments: case studies on children's participation in three European countries', *Childhood* 5(2): 225–239.

Horelli, L. (2007) 'Constructing a theoretical framework for environmental child-friendliness', *Children, Youth and Environment* 17(4): 267–292.

Horelli, L. and Kaaja, M. (2002) 'Opportunities and constraints of "internet-assisted urban planning" with young people', *Journal of Environmental Psychology* 22: 191–200.

Hörschelmann, K. (2009) 'Routes through the city: changing youth identities and spatial practices in Leipzig', in Harutyunyan, A., Hörschelmann, K. and Miles, M. (eds) *Public Spheres after Socialism*, London: Intellect Books.

Hörschelmann, K. and Colls, R. (2009) *Contested Bodies of Childhood and Youth*, London: Palgrave Macmillan.

Hörschelmann, K. and Schäfer, N. (2005) 'Performing the global through the local – young people's practices of identity formation in former east Germany', *Children's Geographies*, 3(2): 219–242.

Hörschelmann, K. and Schäfer, N. (2007) '"Berlin is not a foreign country, stupid!" – Growing up "global" in eastern Germany', *Environment and Planning A* 39/08: 1855–1872.

Horton, J. and Kraftl, P.(2006) 'What else? some more ways of thinking and doing', *Children's Geographies* 4(1): 69–95.

Human Rights Watch (1997) *Juvenile Injustice: Police Abuse and Detention of Street Children in Kenya*, New York: Human Rights Watch.

Iceland, I. J. and Wilkes, R. (2006) 'Does socioeconomic status matter? Race, class, and residential segregation', *Social Problems* 53(2): 248–273.

ILO (2006) *The End of Child Labour: Within Reach*, Global Report under the follow-up to the ILO Declaration on Fundamental Principles and Rights at Work, International Labour Conference, 95th Session 2006, Report I (B), Geneva: ILO.

ILO (2007*) IPEC Action against Child Labour: Highlights 2006*, Geneva: ILO.

Ito, M. and Okabe, D. (2005) 'Intimate connections: contextualizing Japanese youth and mobile messaging', in Harper, R., Palen, L. and Taylor, A. (eds) *The Inside Text: Social Perspectives on SMS in the Mobile Age*, Netherlands: Springer, pp. 127–145.

James, A. (1993) *Childhood Identities: Self and Social Relationships in the Experience of the child*. Edinburgh: Edinburgh University Press.

James, A. and James, A. L. (2004) *Constructing Childhood. Theory, Policy and Social Practice*, London: Palgrave Macmillan.

James, A., Jenks, C. and Prout, A. (2001) *Theorizing Childhood*, Cambridge: Polity.

James, A. and Prout, A. (eds) (1997) *Constructing and Reconstructing Childhood*, Basingstoke: Falmer Press.

James, S. (1990) 'Is there a place for children in geography?' *Area* 22: 278–83.

Jansen, E. (2008) 'Rhymes I wrote', http://www.emileyx.co.za (accessed 15/01/09).

Jeffrey, C. (2008) Generation nowhere: rethinking youth through the lens of unemployed young men. *Progress in Human Geography* 32(6): 739:758.

Jeffrey, C. (2010) Geographies of children and youth I: eroding maps of life. *Progress in Human Geography* 39(4): 496–505.

Jeffrey, C. and McDowell, L. (2004) 'Youth in a comparative perspective: global change, local lives', *Youth and Society* 36: 131–42.

Jenks, C (2005) *Childhood*, London: Routledge.

Jones, M. and Cunningham, C. (1999) 'The expanding worlds of middle-childhood', in Kenworthy Teather, E. (ed.) *Embodied Geographies: Spaces, Bodies and Rites of Passage*, London: Routledge, pp. 27–42.

Jones, O. (1997) 'Little figures, big figures: country childhood stories', in Cloke, J. and Little, J. (eds) *Contested Countryside Cultures: Otherness, Marginalisation and Rurality*, London: Routledge, pp. 158–179.

Jones, O. (2000) 'Melting geography: purity, disorder, childhood and space', in Holloway, S. and Valentine, G. (eds) *Children's Geographies: Playing, Living, Learning*, London: Routledge, pp. 29–47.

Jones, O. (2002) 'Naturally not! Childhood, the urban and romanticism', *Human Ecology Review* 9(2): 17–30.

Kabeer, N. and Mahmud, S. (2009) Imagining the future: children, education and intergenerational transmission of poverty in urban Bangladesh, *IDS Bulletin*, 40(1): 10–21.

Kallio, K. P. (2007) 'Performative bodies, tactical agents and political selves: Rethinking the political geographies of childhood', *Space and Polity* 11: 121–136. Kaplan, C. (1996) *Questions of Travel: Postmodern Discourses of Displacement*, Durham, NC: Duke University Press.

Kallio, K. P. (2008) 'The body as a battlefield: Approaching children's politics', *Geografiska Annaler B* 90: 285–297.

Karsten, L. (1998) 'Growing up in Amsterdam: differentiation and segregation in children's daily lives', *Urban Studies* 35(30): 565–581.

Karsten, L. (2003a) 'Family gentrifiers : challenging the city as a place simultaneously to build a career and to raise children', *Urban Studies* 40(12): 2573–2584.

Karsten, L. (2003b) 'Bleak prospects? Urban planning, family housing and children's outdoor spaces in Amsterdam', *Children's Geographies* 1(2): 295–298.

Kato, Y. (2009) 'Doing consumption and sitting cars: adolescent bodies in suburban commercial spaces', *Children's Geographies* 7(1): 51–66.

Katz, C. (2004) *Growing Up Global. Economic Restructuring and Children's Everyday Lives*, Minneapolis: University of Minnesota Press.

Katz, C. (2006) 'Power, space and terror: social reproduction and the public environment', in Low, S. and Smith, N. (eds) *The Politics of Public Space*, London: Routledge.

Kelley, N. (2006) 'Children's involvement in policy formation', *Children's Geographies* 4(1): 37–44.

Kesby, M. G. (2005) 'Retheorising empowerment-through-participation as a performance in space: beyond tyranny to transformation', *Signs* 30(4): 2037–2065.

Kielland, A. and Tovo, M. (2006) *Children at Work: Child Labour Practices in Africa*, Lynne Rienner Publishers, London.

Kindon, S., Pain, R. and Kesby, M. (2007) 'Introduction: connecting people, participatory action and place', in Kindon, S., Pain, R. and Kesby, M. (eds) *Participatory Action Research Approaches and Methods: Connecting People, Participation and Place*, London: Routledge.

Kingston, B., Wridt, P., Chawla, L., van Vliet, W. and Brink, L. (2007) Creating child friendly cities; the case of Denver, USA, *Municipal Engineer* 160 (Issue ME2): 98–102.

Kjeldgaard, D. and Askegaard, S. (2006) 'The globalization of youth culture: the global youth segment as structures of common difference', *Journal of Consumer Research* 33: 231–247.

Klodawsky, F. (2006) 'Landscapes on the margins: gender and homelessness', *Gender, Place and Culture* 13(4): 365–381.

Knox, P. and Pinch, S. (2006) *Urban Social Geography. An Introduction*, London: Pearson, Prentice Hall.

Kofman, E. and Lebas, E. (eds) (1996) *Writings on Cities. Henri Lefebvre*, Oxford: Blackwell.

Kong, L. (2000) 'Nature's dangers, nature's pleasures: urban children and the natural world', in Holloway, S. and Valentine, G. (eds) *Children's Geographies: Living, Playing, Learning,* London: Routledge, pp. 257–271.

Kong, L., Yuen, B., Sodhi, N. and Briffett, C. (1999) 'The construction and experience of nature: perspectives of urban youths', *Tijdschrift voor Economische en Sociale Geografie*, 90(1): 3–16.

Korpela, K. (1989) 'Place-identity as a product of environmental self-regulation', *Environmental Psychology* 9: 241–256.

Krims, A. (2002) 'Rap, race, the 'local', and urban geography in Amsterdam', *Critical Studies: Music, Popular Culture, Identities* 19: 165–179.

Kruger, J. S. and Chawla, L. (2002) '"We know something someone doesn't know": children speak out on local conditions in Johannesburg', *Environment and Urbanization* 14(2): 85–96.

Langevang, T. (2007) 'Movements in time and space: using multiple methods in research with young people in Accra, Ghana', *Children's Geographies* 5(3): 267–282.

Langevang, T. (2008) 'Claiming place: the production of young men's street meeting places in Accra, Ghana', *Geografiska Annaler: Series B, Human Geography* 90(3): 227–242.

Lansky, M. (1997) 'Perspectives: child labour: how the challenge is being met', *International Labour Review* 136(2): 233–257.

Latham, A (2003) 'Research, performance, and doing human geography: some reflections on the diary-photograph, diary-interview method', *Environment and Planning A* 35: 1993–2017.

Lauby, J. and Stark, O. (1988) 'Individual migration as a family strategy: young women in the Philippines', *Population Studies* 42(3): 473–486.

Lees, L. (1994) 'Rethinking gentrification: beyond the positions of economics or culture', *Progress in Human Geography* 15: 321–337.

Lefebvre, H. (1991) *The Production of Space*, Oxford: Blackwell.

Lemanski, C. (2007) 'Global cities in the south: deepening social and spatial polarisation in Cape Town', *Cities* 24(1): 1–15.

Levy, J. M. (2010) *Contemporary Urban Planning*, 9th edition, London: Pearson.

Lewis, J. J., Ronsmans, C., Ezeh. A. and Gregson, S. (2004) 'The population impact of HIV on fertility in sub-Saharan Africa', *AIDS* 18: 35–43.

Ley, D. (1994) 'Gentrification and the politics of the new middle class', *Environment and Planning D: Society and Space* 12: 53–74.

Ley, D. (1996) *The New Middle Class and the Remaking of the Central City*, Oxford: Oxford University Press.

Liechty, M. (1995) 'Media, markets and modernisation: youth identities and the experience of modernity in Kathmandu, Nepal', in Amit-Talai, V. and Wulff, H. (eds) *Youth Cultures: A Cross-cultural Perspective*, London, Routledge, pp. 166–201.

Lloyd-Evans, S. (2002) 'Child Labour', in Desai, V. and Potter, R. B. (eds) *The Companion to Development Studies*, London: Arnold, pp. 215–218.

Lock, T. (2005) 'I-path UK tour', *Document* January/February: 66–73.

Low, S. (2001) 'The edge and the center: gated communities and the discourse of urban fear', *American Anthropologist* 103(1): 45–58.

Low, S. and Smith, N. (eds) (2006) *The Politics of Public Space*, London: Routledge.

Lowndes, V., Pratchett, L. and Stoker, G. (2001) 'Trends in Public Participation: Part 2 – Citizens' Perspectives', *Public Administration* 79(2): 445–455.

Lynch, K. (1977) *Growing Up in Cities*, Cambridge, MA: MIT Press.

Mac an Ghaill, A. (1994) *The Making of Men: Masculinities, Sexualities and Schooling*, Buckingham: OU Press.

Mac an Ghaill, A. (ed.) (1996) *Understanding Masculinities: Social Relations and Cultural Arenas*, Buckingham: OU Press.

Macdonald, N. (2001) *The Graffiti Subculture. Youth, Masculinity and Identity in London and New York*, Basingstoke: Palgrave.

Maffesoli, M. (1996) *The Time of the Tribes: The Decline of Individualism in Mass Society*, London: Sage.

Malone, K. (2002) 'Street like: youth, culture and competing uses of public space', *Environment and Urbanization* 14(2): 157–168.

Malone, K. and Hasluck, L. (1998) 'Geographies of exclusion: young people's perceptions and use of public space', *Family Matters*, 49: 21–26.

Malone, K. and Hasluck, L. (2002) 'Australian youth', in Chawla, L. (ed.) *Growing Up in an Urbanising World*, London: Earthscan, pp. 81–109.

Massey, D. (1984) *Spatial Divisions of Labour: Social Structures and the Geography of Production*, London: Macmillan.

Massey, D. (1994) *Space, Place and Gender*, Cambridge: Polity Press.

Massey, D. (1995a) *Spatial Divisions of Labour: Social Structures and the Geography of Production*, New York: Routledge.

Massey, D. (1995b) *A Place in the World. Places, Cultures and Globalization*, Milton Keynes: OU Press.

Massey, D (1998) 'The spatial construction of youth cultures', in Skelton, T. and Valentine, G. (eds) *Cool Places: Geographies of Youth Cultures*, London: Routledge, 121–129.

Massey, D. (2005) *For Space*, London: Sage.

Massey, D., Allen, J. and Pile, S. (eds) (1999) *City Worlds*, London: Routledge with the Open University.

Massey, D. S. and Eggers, M. (1990) 'The ecology of inequality: minorities and the concentration of poverty, 1970–1980', *American Journal of Sociology* 95(5): 1153ff.

Matchinda, B. (1999) 'The impact of home background on the decision of children to run away: the case of Yaounde city street children in Cameroon', *Child Abuse and Neglect* 23(3): 245–255.

Matthews, H. (1984) 'Environmental cognition of young children: images of journey to school and home area', *Transactions of the Institute of British Geographers* 9: 89–106.

Matthews, H. (2001) 'Participatory structures and the youth of today: engaging those who are hardest to reach', *Ethics, Place and Environment* 4(2): 153–159.

Matthews, H. (2003) 'Children and regeneration: setting an agenda for community participation and integration', *Children and Society* 17: 264–276.

Matthews, H. and Limb, M. (1999) 'Defining an agenda for the geography of children: review and prospect', *Progress in Human Geography* 23(1): 61–90.

Matthews, H., Limb, M. and Taylor, M. (1999) 'Young people's participation and representation in society', *Geoforum* 30: 135–144.

Matthews, H., Limb, M. and Taylor, M. (2000a) 'The "street as thirdspace"', in Holloway, S. and Valentine, G. (eds) *Children's Geographies: Living, Playing, Learning,* London: Routledge, pp. 63–79.

Matthews, H., Taylor, M., Percy-Smith, B. and Limb, M. (2000b) 'The unacceptable flâneur: the shopping mall as a teenage hangout', *Childhood* 7(3): 279–294.

McCormick, J. and Philo, C. (1995) 'Poor places and beyond: summary findings and policy implications', in Philo, C. (ed.) *Off the Map: The Social Geography of Poverty in the UK*, London: Child Poverty Action Group, pp. 1–22.

McDowell, L. (1999) 'City life and difference: negotiating diversity', in Allen, J., Massey, D. and Pryke, M. (eds) *Unsettling Cities: Movement/Settlement*, London and New York: Routledge, pp. 8–140.

McDowell, L. (2002) Transitions to work: masculine identities, youth inequality and labour market. *Gender, Place and Culture* 9: 39–59.

McDowell, L. (2003) Masculine identities and low paid work: young men in urban labour markets. *International Journal of Urban and Regional Research*, 27(4): 828–848.

McDowell, L. (2004) Work, workfare, work/life balance and an ethics of care. *Progress in Human Geography*, 28(2): 145–163.

McDowell, L. and Sharp, J. (eds) (1997) *Space, Gender, Knowledge: Feminist Readings*, London: Arnold.

McKendrick, J. H., Bradford, M. and Fielder, A. (2000a) 'Kid customer? Commercialization of playspace and the commodification of childhood', *Childhood* 7(3): 295–314.

McKendrick, J., Bradford, M. and Fielder, A. (2000b) 'Time for a party!: making sense of the commercialisation of leisure space for children', in Holloway, S. and Valentine, G. (eds) *Children's Geographies: Living, Playing, Learning,* London: Routledge, pp. 100–118.

McRobbie, A. (1993) 'Shut up and dance: youth culture and changing modes of femininity', *Cultural Studies* 7(3): 406–426.

McRobbie, A (1997) 'Bridging the gap: feminism, fashion and consumption', *Feminist Review* 55: 73–89.

Miles, S. (2000) *Youth Lifestyles in a Changing World*, Buckingham: OU Press.

Mintel (2000) Pre-family leisure trends, Leisure Intelligence, January.

Mintel (1998) Nightclubs and discotheques, Leisure Intelligence, September.

Mitchell, D. (2003) *The Right to the City. Social Justice and the Fight for Public Space*, New York: Guildford Press.

Morin, E. (1962) *Signalement af vor tid*, København: Villadsen & Christensen.

Morrow, V. (1999) '"We are people too": children's and young people's perspectives on children's rights and decision-making in England', *The International Journal of Children's Rights* 7: 149–170.

Mugisha, F. (2006) 'School enrolment among urban non-slum, slum and rural children in Kenya: is the urban advantage eroding?', *International Journal of Educational Development*, 26: 471–482.

Mumford, L. (1937) 'What is a city?', *Architectural Record* LXXXII, November.

Mutersbaugh, T. (2002) 'Migration, common property, and communal labor: cultural politics and agency in a Mexican village', *Political Geography* 21: 473–494.

Mynatt, E., Adler, A., Ito, M. and O'Day, V. (1997) 'Network Communities: Something Old, Something New, Something Borrowed…', *Computer Supported Cooperative Work* 6: 1–35.

Nairn, K., Sligio, J. and Freeman, C. (2006) 'Polarizing participation in local government: which young people are included and excluded?', *Children, Youth and Environments* 16(2): 248–271.

Nakajima, I., Keiichi, H. and Yoshii, H. (1999) 'Ido-denwa Riyou no Fukyuu to sono Shakaiteki Imi' (Diffusion of Cellular Phones and PHS and their Social Meaning), *Tsuushin Gakkai-shi (Journal of Information and Communication Research)* 16(3) (no page numbers).

Närvänen, A-L. and Näsman, E. (2004) 'Childhood as generation or life phase?', *Young: Nordic Journal of Youth Research* 12(1), 71–91.

NCCP (2007) *Basic Facts About Low-Income Children: Birth to Age 18*, National Center for Children in Poverty, September.

Nilan, P. (ed.) (2006) *Hybrid Youth? Hybrid Identities, Plural Worlds*, London: Routledge.

Noack, P. and Silbereisen, R. (1988) 'Adolescent development and choice of leisure settings', *Children's Environments Quarterly* 5: 25–33.

O'Brien, M. (2003) 'Regenerating children's neighbourhoods: what do children want?', in Christensen, P. and O'Brien, M. (eds) *Children in the City. Home, Neighbourhood and Community*, London: Routledge, pp. 142–161.

Orellana, M. F., Thorne, B., Chee, A. and Lam, W. S. E. (2001) 'Transnational childhoods: the participation of children in processes of family migration', *Social Problems* 48(4): 572–591.

Oswalt, P. (ed.) (2004) *Schrumpfende Städte*, Ostfildern-Ruit: Hatje Crantz.

Ovacharova, L. N. and Popova, D. O. (2005) *Child Poverty in Russia*, Moscow: UNICEF.

Owens, P. (1988) 'Natural landscapes, gathering places, and prospect refuges: characteristics of outdoor places valued by teens', *Children's Environments Quarterly* 5: 17–24.

Owens, P. (1994) 'Teen places in Sunshine, Australia: then and now', *Children's Environments* 11: 292–299.

Pain, R (2004) 'Social geography: participatory research', *Progress in Human Geography* 28(5): 652–663.

Pain, R. and Smith, S. (eds) (2008) *Fear: Critical Geopolitics and Everyday Life (Re-materialising Cultural Geography)*, Aldershot: Ashgate.

Payne, L. (2003) 'Anti-social behaviour', *Children & Society* 17: 321–324.

Peake, L. (1999) 'Toward a social geography of the city: race and dimensions of urban poverty in women's lives', *Journal of Urban Affairs* 19(3): 335–361.

Peck, J. and Tickell, A. (2002) 'Neoliberalizing space', in Brenner, N. and Theodore, N. (eds) *Spaces of Neoliberalism: Urban Restructuring in North America and Western Europe*, Malden, MA: Oxford's Blackwell Press, pp. 33–57.

Percy-Smith, B. (2002) 'Contested worlds: constraints and opportunities in city and suburban environments in an English Midlands town', in Chawla, L. (ed.) *Growing up in an Urbanizing World*, London: Earthscan.

Percy-Smith, B. (2006) 'From consultation to social learning in community participation with young people', *Children, Youth and Environments* 16(2): 153–179.

Phillips, S. A. (1999) *Wallbangin'. Graffiti and Gangs in L.A.*, Chicago: University of Chicago Press.

Philo, C. (2003) '"To go back up the side hill": memories, imaginations and reveries of childhood', *Children's Geographies* 1(1): 7–23.

Pilkington, H. (2002) 'The dark side of the moon? Global and local horizons', in Pilkington, H., Omel'chenko, E., Flynn, M., Bliudina, U. and Starkova, E. (eds) *Looking West? Cultural Globalization and Russian Youth Cultures*, University Park: Pennsylvania University Press, pp. 133–164.

Pilkington, H., Omel'chenko, E., Flynn, M., Bliudina, U. and Starkova, E. (2002) *Looking West? Cultural Globalization and Russian Youth Cultures*, University Park: Pennsylvania University Press.

Pilkington, H. with Starkova, E. (2002) '"Progressives" and "normals". Strategies for glocal living', in Pilkington, H., Omel'chenko, E., Flynn, M., Bliudina, U. and Starkova, E. (eds) *Looking West? Cultural Globalization and Russian Youth Cultures*, University Park: Pennsylvania University Press, pp. 101–132.

Platt, L. (2002) *Parallel Lives? Poverty Among Ethnic Minority Groups in Britain,* London: Child Poverty Action Group.

Porter, G. and Abane, A. (2008) 'Increasing children's participation in African transport planning: reflections on methodological issues in a child-centred research project', *Children's Geographies* 6(2): 151–167.

Postman, N. (1994) *The Disappearance of Childhood*, New York: Vintage Books.

Potter, R. B. and Wilson, M. (1991) 'Indigenous environmental learning in a small developing country: adolescents in Barbados, West Indies', *Singapore Journal of Tropical Geography* 11: 56–67.

Pow, C. P. and Kong, L. (2007) 'Marketing the Chinese dream home: gated communities and representations of the good life in (post-)socialist Shanghai', *Urban Geography* 28(2): 129–159.

Pritchard, G. (2009) 'Cultural imperialism, Americanisation and Cape Town hip-hop culture: A discussion piece', *Social Dynamics* 35(1): 51–55.

Prokop, G. (1984) *Detektiv Pinky*, Berlin: Der Kinderbuchverlag Berlin.

Prout, A. (2000a) 'Childhood bodies: construction, agency and hybridity', in Prout, A. (ed.) *The Body, Childhood and Society*, London: Macmillan, pp. 1–18.

Prout, A. (2000b) 'Children's participation: control and self-realisation in British late modernity', *Children and Society* 14: 304–315.

Punch, S. (2000) 'Children's strategies for creating playspaces', in Holloway, S. and Valentine, G. (eds) *Children's Geographies. Playing, Living, Learning*, London: Routledge, pp. 48–62.

Punch, S. (2001) 'Household division of labour: generation, gender, age, birth order and sibling composition', *Work, Employment and Society* 15: 803–823.

Punch, S. (2002) 'Youth transitions and interdependent adult-child relations in rural Bolivia', *Journal of Rural Studies* 18: 123–133.

Punch, S. (2007) 'Negotiating migrant identities: young people in Bolivia and Argentina', *Children's Geographies* 5(1–2): 95–113.

Qvortrup, J. (1994) 'Childhood matters: an introduction', in Qvortrup, J., Bardy, M., Spitta, G. and Wintersberger, H. (eds) *Childhood Matters: Social Theory, Practice and Politics*, Aldershot: Avebury, pp. 1–23.

Qvortrup, J. (1999) *Childhood and Societal Macrostructures. Childhood Exclusion by Default*, Odense: Odense University Press.

Qvortrup, J. (2005) 'Varieties of childhood', in Qvortrup, J. (ed.) *Studies in Modern Childhood: Society, Agency, Culture*, Basingstoke: Macmillan, pp. 1–20.

Racelis, M. and Aguirre, A. D. M. (2002) 'Child rights for the urban poor children in child friendly Phillipine cities: views from the community', *Environment and Urbanization* 14(2): 97–113.

Rahn, J. (2002) *Painting Without Permission. Hip-Hop Graffiti Subculture*, Westport: Bergin and Garvey.

Reay, D. and Lucey, H. (2000) '"I don't really like it here but I don't want to be anywhere else": children and inner city council estates', *Antipode*, 32(4): 410–428.

Respect Task Force (2006) *Respect. Give Respect. Get Respect*, London: Home Office.

Riggio, E. (2002) 'Child friendly cities: good governance in the best interests of the child', *Environment and Urbanization* 14(2): 45–58.

Robertson, R. (1992) *Globalization: Social Theory and Global Culture*, London: Sage.

Robson, E. (2000) 'Invisible carers: young people in Zimbabwe's home-based healthcare', *Area* 32(1): 59–69.

Robson, E. (2004) 'Hidden child workers: young carers in Zimbabwe', *Antipode* 36: 227–248.

Robson, E. (2009) 'Children's bodies: working and caring in Sub-Saharan Africa', in Hörschelmann, K. and Colls, R. (eds) *Contested Bodies of Childhood and Youth*, Basingstoke: Palgrave, pp. 148–162.

Rogers, P. (2006) 'Young people's participation in the renaissance of public space – a case study on Newcastle upon Tyne, UK', *Children, Youth and Environments* 16(2): 105–126.

Rose, N. (1990) *Governing the Soul*, London: Routledge.

Ruddick, S. (1996) *Young and Homeless in Hollywood: Mapping Social Identities*, Routledge, New York.

Ruddick, S. (2007) 'At the horizons of the subject: neo-liberalism, neo-conservatism and the rights of the child part two: parent, caregiver, state', *Gender, Place and Culture* 14(6): 627–640.

Saldanha, A. (2002) 'Music, space, identity: geographies of youth culture in Bangalore', *Cultural Studies* 16(3): 337–350.

Sassen, S. (2000) *Cities in a World Economy*, Thousand Oaks: Pine Forge.

Sassen, S. (2001) *The Global City: New York, London, Tokyo*, Princeton: Princeton University Press.

Schäfer, N. and Yarwood, R. (2008) 'Involving young people as researchers: uncovering multiple power relations among youths', *Children's Geographies* 6(2): 121–135.

Scheper-Hughes, N. and Hoffman, D. (1998) 'Brazilian apartheid: street kids and the struggle for urban space', in Scheper-Hughes, N. and Sargent, C. (eds) *Small Wars: The Cultural Politics of Childhood*, Berkeley: University of California Press, pp. 352–388.

Schütz, A. (1972) *The Phenomenology of the Social World*, trans. G. Walsh and F. Lehnert, London: Heinemann.

Scott, A., Agnew, J., Soja, E. and Storper, M. (2001) 'Global city-regions: an overview', in Scott, A. (ed.) *Global City-Regions: Trends, Theory, Policy*, Oxford: Oxford University Press, pp. 11–30.

Searle, B. (2008) *Well-being: In Search of a Good Life?* Bristol: Policy Press.

Sen, R. and Goldbart, J. (2005) 'Partnership in Action: introducing family-based intervention for children with disability in urban slums of Kolkata, India', *International Journal of Disability, Development and Education* 52(4): 275–311.

Sener, T. (2006) 'The children and architecture project in Turkey', *Children, Youth and Environments* 16(2): 191–206.

Sennett, R. (2005) *The Culture of the New Capitalism*, New Haven, CT: Yale University Press.

Shields, R. (1999) *Lefebvre, Love and Struggle. Spatial Dialectics*, London: Routledge.

Short, J. R. (1991) *Imagined Country: Society, Culture and Environment*, London: Routledge.

Sibalis, M. (2004) 'Urban space and homosexuality: the example of the Marais, Paris "Gay Ghetto"', *Urban Studies* 41(9): 1739–1758.

Sibley, D. (1991) 'Children's geographies: some problems of representation', *Area* 23: 269–270.

Sibley, D. (1995) *Geographies of Exclusion*, London: Routledge.

Silva, A. (1986) *Una ciudad imaginada: graffiti, expresión urbana*, Bogotá: Universidad Nacional de Colombia.

Silvey, R. (2001) 'Migration under crisis: household safety nets in Indonesia's economic collapse', *Geoforum* 32(1): 33–45.

Sinclair, R. (2004) 'Participation in practice: making it meaningful, effective and sustainable', *Children and Society* 18: 106–118.

Skelton, T. (2007) 'Children, young people, UNICEF and participation', *Children's Geographies*, 5(1): 165–181.

Skelton, T. (2010) 'Taking young people as political actors seriously: opening the borders of political geography', *Area* 42(2): 145–151.

Skelton, T. and Allen, T. (1999) *Culture and Global Change*, London: Routledge.

Skelton, T. and Valentine, G. (1998) *Cool Places: Geographies of Youth Cultures*, London: Routledge.

Skelton, T. and Valentine, G. (2003) '"It feels like being Deaf is normal": an exploration into the complexities of defining D/deafness and young D/deaf people's identities', *Canadian Geographer* 47(4): 451–466.

Smith, D. (2005) 'Studentification: the gentrification factory?', in Atkinson, R. and Bridge, G. (eds) *Gentrification in a Global Context: The New Urban Colonialism*, London: Routledge, pp. 72–89.

Smith, D. and Holt, L. (2007) 'Studentification and "apprentice" gentrifiers within Britain's provincial towns and cities: extending the meaning of gentrification', *Environment and Planning A* 39: 142–161.

Smith, F. and Barker, J. (2000a) '"Out of school", in school: a social geography of out of school childcare', in Holloway, S. and Valentine, G. (eds) *Children's Geographies: Playing, Living, Learning,* London: Routledge, pp. 245–256.

Smith, F. and Barker, J. (2000b) 'Contested spaces: children's experiences of out of school care in England and Wales', *Childhood: A Global Journal of Child Research*, special issue on spaces for children, 7(3): 315–333.

Smith J. (1999) 'Youth homelessness in the UK. A European perspective', *Habitat International* 23(1): 63–77.

Smith, N. (1996) *The New Urban Frontier: Gentrification and the Revanchist City*, New York: Routledge.

Smith, S. J. (1988) 'Political interpretations of "racial segregation" in Britain', *Environment and Planning D: Society and Space* 6: 423–444.

Sommer, B. (1990) 'Favourite places of Estonian adolescents', *Children's Environments Quarterly* 7: 32–36.

Stevenson, S. (2001) 'Street children in Moscow: using and creating social capital', *The Sociological Review*: 530–547.

Sumka, H. J. (1979) 'Neighborhood revitalization and displacement: a review of the evidence', *Journal of the American Planning Association* 45: 480–487.

Swanson, K. (2005) *Begging for Dollars in Gringopampa: Geographies of Gender, Race Ethnicity and Childhood in the Ecuadorian Andes*, PhD thesis, Graduate Department of Geography, University of Toronto.

Swanson, K. (2010) Begging as a path to progress: indigenous women and children and the struggle for Ecuador's urban spaces: indigenous women and children and the struggle for Ecuador's urban spaces, *Geographies of Justice and Social Transformation Series*. Athens: University of Georgia Press.

Tacon, P. (1982) 'Carlinhos: the hard gloss of city polish', *UNICEF News*, 111(1): 4–6.

Thorne, B. (1987) Re-visioning women and social change where are the children? *Gender & Society* 1(1): 85–109.

Thornton, S. (1995) *Club Cultures*, Cambridge: Polity Press.

Tomlinson, J. (1999) *Globalization and Culture*, Oxford: Polity Press.

Townsend, A. (1979) *Poverty in the United Kingdom: A Survey of Household Resources and Living Standards*, London: Penguin.

Tucker, F. (2003) 'Sameness or difference? Exploring girls' use of recreational spaces', *Children's Geographies* 1(1): 111–124.

UN-HABITAT (2008) *State of the World's Cities 2008/2009: Harmonious Cities*, London: Earthscan.

UNICEF (2000) *A League Table of Child Poverty in Rich Nations*, Innocenti Report Crad 1, Florence.

UNICEF (2007) *Child Poverty in Perspective: An Overview of Child-Wellbeing in Rich Countries. A comprehensive assessment of the lives and well-being of children and adolescents in the economically advanced nations*, Innocenti Report Card 7, Florence.

United Nations Center for Human Settlements (1996) *An Urbanizing World: Global Report on Human Settlements*, Oxford: Oxford University Press for HABITAT.

Valentine, G. (1995) '"Out and about": a geography of lesbian communities', *International Journal of Urban and Regional Research* 19(3): 96–111.

Valentine, G. (1996a) 'Angels and devils: moral landscapes of childhood', *Environment and Planning D: Society and Space* 14: 581–599.

Valentine, G (1996b) 'Children should be seen and not heard: the production and transgression of adults' public space', *Urban Geography* 17: 205–220.

Valentine, G. (1997a) '"Oh yes I can." "Oh no you can't": children and parents' understandings of kids' competence to negotiate public space safely', *Antipode* 29(1): 65–89.

Valentine, G. (1997b) '"My son's a bit dizzy." "My wife's a bit soft": gender, children and cultures of parenting', *Gender, Place and Culture* 4(1): 37–62.

Valentine, G. (1999a) 'Exploring children and young people's narratives of identity', *Geoforum* 31: 257–267.

Valentine, G (1999b) 'Being seen and heard? The ethical complexities of working with children and young people at home and at school', *Ethics, Place and Environment,* 2(2): 141–155.

Valentine, G. (2004) *Public Space and the Culture of Childhood*, Aldershot: Ashgate.

Valentine G. and McKendrick J. (1997) Children's outdoor play: exploring parental concerns about children's safety and the changing nature of childhood', *Geoforum* 28(2): 219–235.

Valentine, G. and Skelton, T. (2003) 'Finding oneself, losing oneself: the lesbian and gay "scene" as a paradoxical space', *International Journal of Urban and Regional Research* 27(40): 849–866.

Valentine, G., Butler, R. and Skelton, T. (2001) 'The ethical and methodological complexities of doing research with "vulnerable" young people', *Ethics, Place and Environment* 4(2): 119–124.

Valentine, G., Holloway, S. and Jayne, M. (2010) 'Contemporary cultures of abstinence and the nighttime economy: Muslim attitudes towards alcohol and the implications for social cohesion', *Environment and Planning A* 42: 8–22.

Valentine, G., Skelton, T. and Butler, R. (2003) 'Coming out and outcomes: negotiating lesbian and gay identities with, and in, the family', *Environment and Planning D: Society and Space* 21(4): 479–499.

Valentine, G., Skelton, T. and Chambers, D. (1998) 'Cool places. An introduction to youth and youth cultures', in Skelton, T. and Valentine, G. (eds) *Cool Places: Geographies of Youth Cultures*, London: Routledge, pp. 1–34.

van Blerk, L. (2006) 'Diversity and difference in the everyday lives of Ugandan street children', *Social Dynamics* 32(1): 47–74.

van Blerk, L. (2007a) 'Mobile lives: the socio-spatial experiences of young commercial sex workers in Ethiopia', unpublished paper.

van Blerk, L. (2007b) 'AIDS, mobility and commercial sex in Ethiopia: implications for policy', *AIDS Care* 19(1): 79–86.

van Blerk, L. (2008) Poverty, migration and sex work: youth transitions in Ethiopia, *Area* 40(2): 245–253.

van Blerk, L. (2009) 'AIDS, mobility and commercial sex in Ethiopia: implications for policy', in Hörschelmann, K. and Colls, R. (eds) *Contested Bodies of Childhood and Youth*, Basingstoke: Palgrave, pp. 232–246.

van Blerk, L. (2011) 'Negotiating boundaries: bar girls and sex work in Ethiopia', *Gender, Place and Culture* 18(2): 217–233.

van Blerk, L. and Ansell, N. (2006) 'Moving in the wake of AIDS: children's experiences of migration in Southern Africa', *Environment and Planning D: Society and Space,* 24(3): 449–471.

Van der Burgt, D. (2008) 'How children place themselves and others in local space', *Geografiska Annaler: Series B, Human Geography* 90(3): 257–269.

Vanderbeck, R. M. (2007) 'Intergenerational geographies: age relations, segregation and re-engagements', *Geography Compass* 1/2: 200–221.

Vanderbeck, R. and Johnson, J. H. (2000) 'That's the only place where you can hang out: urban young people and the space of the mall', *Urban Geography* 21: 5–25.

Van Roosmalen, E. and Krahn, H. (1996) 'Boundaries of youth', *Youth and Society* 28: 3–39.

Ward, C. (1978/1990) *The Child in the City*. London: Bedford Square Press

Warner, M. E. (2008) *Planning Family Friendly Cities: Context and Opportunities*, Presentation to the American Planning Association, April 29, 2008, Las Vegas, NV.

Watson, V. and McCarthy, M. (1998) 'Rental housing policy and the role of the household rental housing sector: evidence from South Africa', *Habitat International* 22(1), 49–56.

Weller, S. and Bruegel, I. (2009) 'Children's "place" in the development of neighbourhood social capital', *Urban Studies* 46(3): 629–643.

White, P. (1999) 'Ethnicity, racialization and citizenship as divisive elements in Europe', in Hudson, R. and Williams, A. M. (eds) *Divided Europe: Society and Territory*, London: Sage, pp. 210–230.

Wilhjelm, H. (2002) 'Large but not unlimited freedom in a Nordic city', in Chawla, L. (ed.) *Growing Up in an Urbanising World*, London: Earthscan, pp. 161–182.

Wilkinson, P. (2000) 'City profile: Cape Town', *Cities* 17(3): 195–205.

Willis, P. (1977) *Learning to Labour*, Aldershot: Gower.

Winchester, H. (1991) 'The geography of children', *Area* 23: 357–360.

Winterbottom, D. (2008) 'Garbage to garden: developing a safe, nurturing and therapeutic environment for the children of the garbage pickers utilizing an academic design/build service learning model', *Children, Youth and Environments* 18(1): 435–455.

Woolley, H. (2007) 'Where do the children play? How policies can influence practice', *Proceedings of the Institution of Civil Engineers: Municipal Engineer*, 160(ME2): 89–95.

Wyness, M. (2000) *Contesting Childhood*, London: The Falmer Press.

Young, L. (2003) 'The place of street children in Kampala, Uganda: marginalisation, resistance and acceptance in the urban environment', *Environment and Planning D: Society and Space* 21(5): 607–628.

Young, L. (2004) 'Journeys to the street: the complex migration geographies of Ugandan street children', *Geoforum* 35(4): 471–488.

Young, L. and Ansell, N. (2003) 'Fluid households, complex families: the impacts of children's migration as a response to HIV/AIDS in southern Africa', *The Professional Geographer* 55: 464–479.

Young, L. and Barrett, H. (2001) 'Adapting visual methods: action research with Kampala street children', *Area* 33(2): 141–152.

Zelizer, V. A. (1994) *Pricing the Priceless Child: The Changing Social Value of Children*, Princeton, NJ: Princeton University Press.

Zhou, M. (1997) 'Growing up American: the challenge confronting immigrant children and children of immigrants', *Annual Review of Sociology* 23: 63–95.

Zouev, A. (ed.) (1998) *Generation in Jeopardy: Children in Central and Eastern Europe and the Former Soviet Union,* UNICEF, London, M.E. Sharpe.

Zukin, S. and Smith Maguire, J. (2004) 'Consumers and consumption', *Annual Review of Sociology* 30: 173–197.

Index

Page numbers in *italic* denote an illustration/table